SCIENCE
AND
SOCIETY

SCIENCE
AND
SOCIETY
IN THE SIXTEENTH AND
SEVENTEENTH CENTURIES

ALAN G. R. SMITH

with 135 illustrations, 16 in color

HARCOURT BRACE JOVANOVICH, INC.

Frontispiece
1   Examining the moon through a telescope:
panel of a triptych by Donato Creti.

© 1972 THAMES AND HUDSON LTD, LONDON

First American edition 1972

ISBN 0–15–578399–8

Library of Congress Catalog Card Number: 75-187701

PRINTED AND BOUND IN GREAT BRITAIN BY JARROLD AND SONS LTD, NORWICH

CONTENTS

The scientific advances of the sixteenth and seventeenth centuries rank among the greatest creative changes in world history: they reshaped man's view of the universe and of his own place in it. At first the new ideas were understood only by the educated minority of the western European population, but it was that minority which was the dynamic element in the society of the time, exercising a profound influence on religion, government, economic life and literature, all of which were affected by the new science. By the end of the seventeenth century educated men were living in a different mental world from their predecessors in 1500.

Moreover, the seventeenth-century scientific revolution changed the whole course of European civilization, not only because of its effects in the period up to 1700, important though these were, but also because of its crucial links with the spectacular advances of modern science. There is no doubt about the role of nineteenth- and twentieth-century science in shaping the whole economic and social life of the contemporary world, which is founded upon the technological achievements made possible by the fundamental scientific discoveries of the last hundred years. The scientists who made these discoveries, which taken together constitute a second scientific revolution, owed one debt above all others to their seventeenth-century predecessors, who created the first scientific revolution. Seventeenth-century scientists developed a new methodology for investigating the phenomena of nature which their successors only *applied*, though with great sophistication and outstanding success. As a result, there have been, in the twentieth century, advances in physics and astronomy – most notably, perhaps, the development of the relativity and quantum theories – which have created a picture of the workings of the universe which is almost as much of an advance on the ideas of 1700 as these were on the concepts of 1500.

This book is an attempt to discuss the significance of those scientific discoveries which later historians have recognized as most important in establishing the new ideas about man and the universe which were emerging by 1700. It therefore omits areas of sixteenth- and seventeenth-century science, such as alchemy and astrology, which were of major concern at the time but have since come to be regarded as 'unscientific'.

I wish to thank the General Editor, Professor Geoffrey Barraclough, and my colleagues, Peter Parish, Geoffrey Finlayson and Brian Dietz, for their helpful suggestions for improving the text. I would also like to thank Stanley Baron, Christopher Pick, Kathryn Freeman and Tessa Campbell of Thames and Hudson for their help.

ALAN G. R. SMITH

University of Glasgow
August 1971

I SCIENCE AND SOCIETY IN 1500

In the later Middle Ages the dominant philosophical influence in western Europe was that of Aristotle, who lived in the fourth century BC and whose work was reconciled with Christian theology in the thirteenth century in the great synthesis of St Thomas Aquinas. In 1500, therefore, men's cosmological ideas were founded upon a mixture of Christian and Aristotelian thought. The earth, it was believed, was motionless at the centre of the universe. Around it were a number of moving spheres, made of a crystalline substance, which circled the earth and filled the whole of space. Aristotle had envisaged eight spheres, but later astronomers added two others, so that in 1500 most models of the universe consisted of ten spheres. Eight of these, working outwards from the earth, contained the moon, Mercury, Venus, the sun, Mars, Jupiter, Saturn and the fixed stars. The ninth and tenth spheres carried no heavenly bodies and their movements were used to account for slight alterations which took place over the centuries in the positions of the fixed stars as seen from the earth. These changes were actually produced by a slow alteration of the direction in space of the earth's axis, but in the Middle Ages, when the earth was believed to be at rest, the observed phenomena could only be explained by adding two extra spheres to the eight which carried the heavenly bodies. Beyond the tenth sphere was heaven, containing the throne of God and the home of the elect. The spheres were just thick enough for their planets to be at the outer surface when furthest away from the earth and at the inner surface when nearest to the earth. The spheres themselves were kept in motion by angels.

This picture of the universe, which seems so strange and ridiculous to modern man, fitted in very well with both the

2, 3 In the universe pictured by medieval thinkers, all things in motion required the constant attention of a mover; hence the angels (above) needed to turn the planetary spheres. In the Ptolemaic system, which remained dominant until the time of Copernicus, the spheres carried the heavenly bodies round the earth, which lay motionless at the centre of the universe. The tenth sphere was the *Primum Mobile*, the 'first mover', which turned the skies round every twenty-four hours to produce the alternation of day and night.

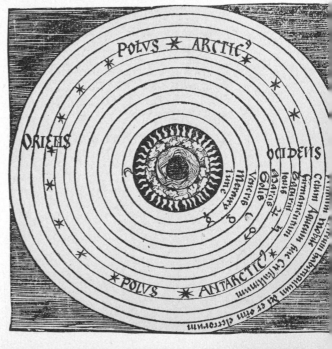

theological ideas and 'common sense' of the time. In theology it provided a specific home for God and for the souls of redeemed men and gave the angels a role in working the mechanisms of the heavens. But it also fitted in, as well as any system could, with men's practical view of the universe as they looked up at the night sky from their villages and little towns. It seemed very obvious to them that the huge solid earth was motionless and that the light ethereal heavenly bodies circled around it.

The idea of concentric crystalline spheres carrying the planets, which provided men's general picture of the universe, was not, however, adequate for professional astronomers, who wanted to know the precise paths of the planets across the sky. The continual movements of the seven planetary spheres around the earth were not sufficient to explain these, and astronomers had to resort to a series of elaborate calculating devices to determine them. In 1500 all such calculations were still based essentially on the work of Claudius Ptolemy, the greatest astronomer of antiquity, who lived in Alexandria in the second

4 Pages from a sixteenth-century edition of Ptolemy's *Almagest*, showing the intricate mathematical treatment used to account for the complex planetary motions.

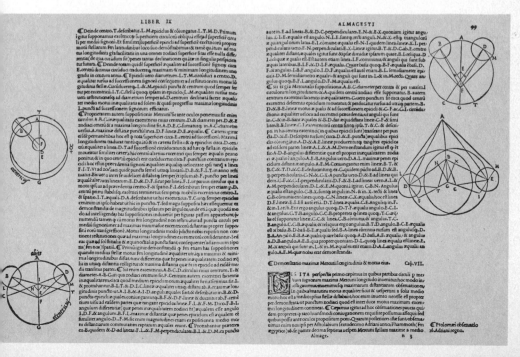

century A D. In his *Almagest* (from the Arabic word for 'greatest', used to distinguish the book from more elementary treatises) Ptolemy discussed in detail the mathematical devices which he used to determine the exact paths of the sun, moon and planets across the sky. These devices included a new invention of his own, the equant, and it was Copernicus's dissatisfaction with this device which was largely responsible for his developing in the sixteenth century the new astronomical system which replaced the earth by the sun as the centre of the universe and constituted the first great step in the scientific revolution.

According to Aristotelian theory the heavenly bodies all moved in exactly circular orbits and each moved at the same uniform speed throughout its entire orbit. Professional astronomers during the whole period after Aristotle up to and including Copernicus in the sixteenth century found it very difficult to reconcile these ideas with the observed movements of the planets. This was hardly surprising – as is now known, the planets neither move in circular orbits nor at uniform speeds throughout their orbits! – but the influence of Aristotle was so strong that astronomers went on trying to harmonize his ideas with the physical reality of the planets' movements. The mathematical devices which appear in the work of Ptolemy and other later astronomers were all designed to achieve this objective.

Ptolemy introduced the equant to meet a specific difficulty. Using older geometrical constructions known as epicycles and eccentrics he had succeeded in constructing a series of circular orbits for the sun, moon and planets. Having done so, however, he found that the heavenly bodies were apparently moving at non-uniform rates in their orbits. This, of course, was not allowed, as it violated one of Aristotle's axioms, and Ptolemy 'solved' the problem by supposing that a planet's rotation was uniform as measured not from the centre of its orbit – the earth – but from an arbitrarily fixed point in space – the equant.

In fact, Ptolemy's whole system was a series of calculations designed to account in different ways for different aspects of a

5 Ptolemy using a
quadrant to measure the
altitude of the moon; he
is guided by the muse of
astronomy: illustration
from an early sixteenth-
century manuscript.

planet's motions: one set of calculations was used to explain a
planet's uniform speed around the earth, a quite different set
to explain its circular motion around the earth. He did not
provide a single set of calculations capable of explaining all the
motions of any one planet at the same time, let alone all the
motions of all the planets at the same time. However, this did
not unduly worry him or his successors. Professional astronomers were simply concerned to 'save the appearances'; that
is, to provide calculating devices which would determine the
tracks of the planets as exactly as possible, while at the same
time conforming to Aristotelian ideas of uniform circular
motion. By 1500, in fact, astronomers had made no real progress since Ptolemy's day. They still had the same ideas and
were still concerned with the same problems.

One of the most notable results of this preoccupation with
calculating devices was a divorce between observational

13

6 The Greek philosopher, Aristotle (384–322 BC), whose scientific and cosmological ideas dominated European science for almost two thousand years.

astronomy and physical reality. Astronomers, like the other men of their time, generally accepted some version of Aristotle's picture of concentric crystalline spheres when they thought about the physical structure of the universe, but in their professional work they were almost totally unconcerned with physical reality – they simply wanted to adjust and rearrange the Ptolemaic epicycles, eccentrics and equants in such a way that these would account for the planetary motions with the greatest possible precision. It did not matter to them that these devices had no existence in the real physical universe. It was left to Copernicus to begin to mend this gaping breach between astronomy and physics, though that work was not finally completed until the time of Newton.

The profound influence of Aristotle in the fields of cosmology and astronomy was paralleled by his equally great authority in physics. His views on matter and on the motions of bodies still dominated scientific thought in 1500, though his mechanics – his ideas about motion – had been considerably modified by medieval scholars. Aristotle made a fundamental distinction as far as matter and motion were concerned between the sublunary and superlunary worlds – the areas below and above the sphere of the moon. In the superlunary universe matter was incorruptible and consisted of the 'quintessence' or fifth essence,

14

7 Medieval scientists – clerk, astronomer (with astrolabe) and mathematician – portrayed in a thirteenth-century manuscript. ▶

so called to distinguish it from the four elements of the sub-lunary sphere. The incorruptible heavenly bodies composed of the quintessence moved around the earth in circles. This was because circular movement was, Aristotle believed, the most 'perfect' kind of motion and thus appropriate for the 'perfect' heavenly bodies. To quote his own words: 'The unceasing movement of the heavens is perfectly understandable: everything ceases to move when it comes to its natural destination, but for the body whose natural path is a circle, every destination is a fresh starting point.'

In the sublunary sphere – notably on the earth itself – matter consisted of four elements: fire, air, water and earth. These, unlike the unchanging material of the heavens, were 'corruptible' or subject to change, and their 'natural' motion was vertical, either downwards, in the case of earth and water, or upwards, in the case of air and fire. Natural motion did not

8 Archimedes, the greatest mathematician of antiquity, standing on the earth, surrounded by the other three elements. Beyond the outer sphere – of fire – lies the region of the sun, stars and planets, composed of incorruptible matter.

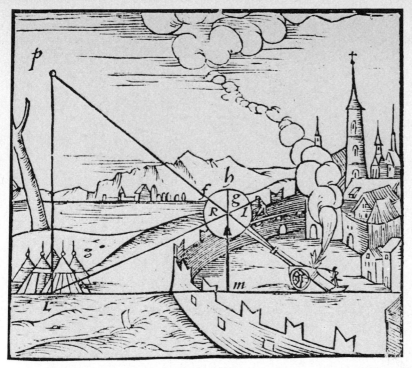

9 The Aristotelian theory of motion: the projectile moves in a straight line from the mouth of the cannon, and then it drops vertically to earth; later Galileo discovered that the true path was that of a parabola.

always happen, of course, partly because the elements were often mixed together in different proportions and partly because they were subjected to 'violent' motions, that is to say motions brought about by some outside agent – man, for example.

Aristotle's ideas about motion are of the utmost importance, because the changeover – in the broadest sense – from his beliefs on that subject to those of Galileo, Descartes and Newton was one of the central elements, perhaps the central element, in the interlocking series of discoveries and changed assumptions that made up the scientific revolution. Aristotle's view of violent motion, just like his view of the universe itself, was essentially a common-sense one. Its central assumption was that a body would remain in movement only as long as a mover was actually in contact with it. Immediately the mover stopped, the movement ceased, and the body came to rest or fell to the ground. Moreover, he argued that a uniform force applied to a body moved it at a constant speed. On the face of it, much of this seems a reasonable picture of the movement of bodies in terms of our everyday experience, but it contains basic fallacies, and the scientific revolution could not have taken place unless these had been overcome.

These ideas about motion did lead Aristotle and his followers into certain difficulties; two of these in particular merit some attention, as they preoccupied scholars in the Middle Ages and led to significant advances beyond the Aristotelian position. One concerned projectiles, the other freely falling bodies. On Aristotle's arguments, as we have followed them so far, projectiles, for example arrows from a bow or (anachronistically) balls from a cannon, ought to fall to the ground the moment they leave the bowstring or the cannon-mouth, because direct contact with the original mover is then broken. Aristotle explained the continued movement of the arrows or cannon-balls by arguing that they caused a disturbance in the air, which then rushed round behind them and forced them on their way. It was not a very satisfactory solution, as the air, which was normally regarded as being a force resistant to violent motion, had to do duty here as an aid to it, but it was the best that Aristotle and his followers could do, and it was not really improved upon in a systematic way until the development of the idea of impetus in the fourteenth century.

As far as falling bodies were concerned, it was obvious to any observer that these moved at a constantly *accelerating* speed. As, however, the force acting upon such bodies seemed the same throughout their period of fall, this did not fit in with Aristotle's fundamental idea that a uniform force produced a uniform speed in the body on which it acted. The Aristotelians had two different arguments to explain this discrepancy. They either insisted that a body moved more quickly as it approached the earth because it was drawing nearer to its 'natural' resting-place in the universe, or else they claimed that as it neared the earth the weight of the atmosphere pressing upon it grew greater and thus imparted a greater force, which in turn produced a greater speed. Those explanations, like the one offered for the continued movement of projectiles, were not very convincing, and yet, despite the challenges which came to them from the work of the medieval schoolmen, they still persisted in 1500. This was far less because of their own merit than because they formed part of the great Aristotelian chain of ideas about heaven and earth. It was difficult to reject part of that chain without bringing the whole into jeopardy, and men who in 1500 could conceive of no alternative framework for the universe were naturally reluctant to do this.

The Aristotelian universe of concentric crystalline spheres with the earth at its centre was thus, in 1500, a generally accepted part of the mental world of European men, who had equally decided views about their own central importance in the scheme of creation. They saw themselves as the vital link in a great hierarchy of living creatures which ranged in dignity from the mightiest angels at the foot of God's throne in heaven to the meanest life forms on earth. This idea of a 'great chain of being', which dated from the time of Aristotle's teacher Plato and continued to play an important role in European thought until the end of the eighteenth century, bore witness to men's belief in an ordered hierarchy in the universe. Some aspects of the concept were expressed by Shakespeare in Ulysses' famous speech about degree in *Troilus and Cressida*; others were well summarized by an Italian count, Hannibal Romei. He did not publish his *Courtier's Academy* until 1546 but the ideas in it reflected the beliefs of 1500.

The most excellent and great God, having with all beauty bedecked the celestial regions with angelic spirits, furnishing the heavenly spheres with souls eternal and having replenished this inferior part with all manner of plants, herbs and living creatures, the divine majesty, desirous to have an artificer who might consider the reason of so high a work, admire the greatness and love the beauty thereof, in the end made man, being of all worldly creatures the most miraculous. But this divine workman, having before the creation of man dispensed proportionably of his treasures to all creatures and every kind of living thing, prescribing unto them infallible laws, as to plants nourishment, to living creatures sense, and to angels understanding, and doubting with what manner of life he should adorn this his new heir, this divine artificer in the end determined to make him, unto whom he could not assign anything in proper, partaker of all that which the others enjoyed, but in particular. Whereupon, calling unto him he said: 'Live, O Adam, in what life pleaseth thee best and take unto thyself those gifts which thou esteemest most dear.' From this so liberal a grant had our free will its original, so that it is in our power to live like a plant, living creature, like a man, and lastly like an angel; for if a man addict himself only to feeding and nourishment he becometh a plant, if to things sensual he is a brute beast, if to things reasonable and civil he groweth a celestial creature; but if he exalt the beautiful gift of his mind to things invisible and divine he transformeth himself into an angel and to conclude, becometh the son of God.

Here is expressed not only the conception of man's nodal position in the chain of being, but also the idea that he could by his own efforts transform himself into one of the elect, a 'son of God'. Man's life on earth, in fact, was regarded essentially as a preparation for existence after death. What mattered was not really happiness in this life, but virtuous and godly conduct which would lead to a place in heaven; though such salvation could only be certainly secured through the mediation of the hierarchically organized Catholic Church, which epitomized on a smaller scale the hierarchically organized universe.

These philosophical and theological views of man's place in

creation and his role on earth were founded upon a combination of Christian doctrine and Platonic theory, in which the former element played the larger part. But in ideas about the structure and functioning of the human body – in other words in the more mundane matters of biology and medicine – it was the views of the ancients that were predominant. Here too, just as in astronomy and physics, Aristotle had considerable prestige, but the leading authority was undoubtedly the physician Galen, born in Asia Minor in the second century A D. His ideas on anatomy, physiology and on the treatment of disease still dominated medical thinking in 1500.

Galen made a series of remarkable experiments on both dead and living animals, including apes, but human dissection was not practised in his time, and his reliance on animal anatomy without sufficient checks upon human subjects led him into numerous errors about the human body. For example, the *rete mirabile*, a structure described by Galen as part of the human body, did not in reality exist at all in man: it was a feature of animal anatomy. By the later Middle Ages, however, when human dissection was practised, Galen's authority was so firmly entrenched that professional anatomists did not dream of challenging his views. In 1500 the teaching of anatomy in the European universities was a set ritual. The professor read a Galenic text while a humble demonstrator dissected a human body for purposes of illustration. Clearly, the study of anatomy could not make substantial advances until Galen's views were subjected to specific criticism based on practical experiments, a situation which did not come about until the work of Vesalius in the early sixteenth century.

If Galen's views on anatomy – the arrangement of the organs – were often misleading, his ideas about physiology – the functioning of the body – were even more mistaken. No accurate physiological system could be constructed until the discovery of the circulation of the blood was made in the seventeenth century. In the Middle Ages the deeply entrenched belief that only heavenly matter could move naturally in a circle and

Anathomia Mū
dini Emēdata p
doctoꝛe melerstat

10, 11 Post-mortem examinations were regarded as sinful until well into the Middle Ages. Above: an early representation (*c.* 1300) of a dissection. The surgeon, who ha removed various organs, is being rebuked by a physician and a monk. But by the end of the fifteenth century the Christian Church was more tolerant of the kind of demonstration shown on this title-page, left, from a book on anatomy; an assistant conducts the dissection while the lecturer reads from his copy of Galen.

12 Opposite: a late thirteenth-century illustration of the venous system within the body. According to Galen the venous system was distinct from the arterial, and the blood ebbed and flowed through the body.

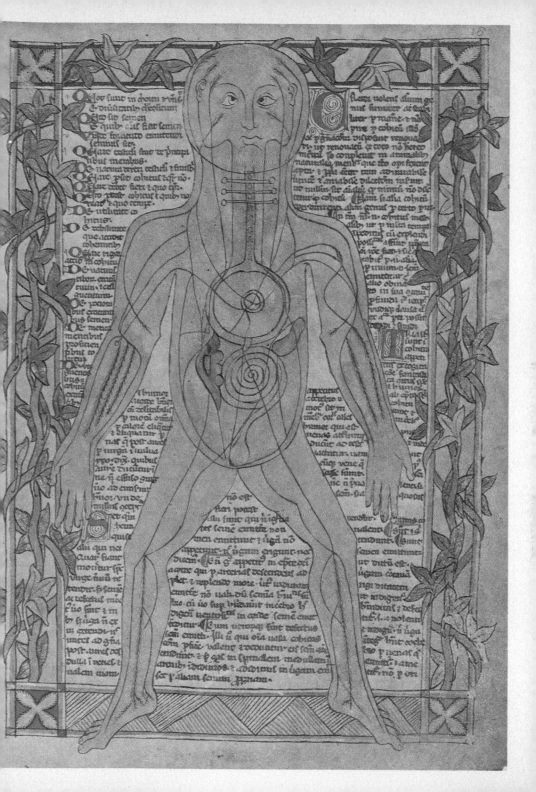

13 The theory of the four bodily humours, popularized by Galen, enjoyed great influence during the Middle Ages and even later. The drawing depicts, clockwise, from top left, the melancholic man (pessimistic), the sanguin (optimistic), the phlegmatic (stolid) and the choleric (excitable).

that all natural motions on earth were rectilinear formed a profound mental barrier to the development of any kind of circulation theory. In these circumstances, Galen's views, which postulated a kind of ebb and flow of the blood upwards and downwards in the body, fitted in admirably with general mental assumptions. According to Galen there were actually two distinct kinds of blood, each of which had its own separate function: a bright red blood, which governed men's muscular activities and flowed upwards and downwards through the arteries, and a dark red blood, which governed the digestive processes and ebbed and flowed in the veins. Here was a further obstacle to any idea of a circulation of the blood; large-scale movement of blood from the arteries to the veins and the veins

24

to the arteries, necessary in any circulation theory, would have required the mixing of what were regarded as quite different fluids, each with its own function to perform.

The medical practitioners of 1500, who so readily accepted Galen's views on physiology and anatomy, also followed his ideas about the treatment of disease. These were based on the doctrine of the four bodily humours, which Galen took over from Aristotle and other thinkers of antiquity. The four humours were blood, which was warm and moist; phlegm, cold and moist; yellow bile, warm and dry; and black bile, cold and dry. Disease, it was believed, was caused by an imbalance of these humours, which resulted in a disturbance of the whole body. The physician's duty was to restore health by restoring the balance of the humours. This idea severely limited the attention paid to diagnosis, as little or no attempt was made to differentiate diseases according to either the organ affected or the external cause of the trouble. Physical examination of patients was not generally considered necessary, and the most common form of diagnosis, 'water-casting' as it was called, consisted of an examination of the patient's urine. Urine was regarded as a filtered overflow of the blood, and the imbalance of the humours could be studied from the quantity and colour of urine samples.

14   Examining a sample of a patient's urine: illustration from a late fifteenth-century book.

15 Bleeding a patient: illustration from a thirteenth-century French manuscript.

Such generally unsound principles of diagnosis were accompanied by primitive treatments. It is true that some of the great variety of herbal medicines used had beneficial effects, but the more drastic forms of remedy, savage purgings and bleedings, only too often resulted in the death of the patient. Clearly, in 1500 men's ideas about the structure and functioning of the body and about its treatment in times of sickness were almost as mistaken as their views about the structure of the universe itself.

The mistaken 'scientific' ideas of 1500 were shared by virtually all sections of society. There was no significant difference between the general assumptions of the highly educated 'scientists' of the universities on the one hand and the illiterate ploughmen of the fields on the other. All accepted the idea of an earth-centred universe dominated by man, and all believed in the fundamentals of a Christian religion which taught that man's life on earth was merely a preparation, comparatively unimportant in itself, for a more glorious afterlife, the way to which had been shown by Christ, the Man-God who had died on the Cross for the redemption of mankind.

Two centuries later these common assumptions no longer existed. By 1700 many members of the educated élites of western Europe had, in the broadest sense, absorbed the implications of the scientific discoveries of the sixteenth and seventeenth centuries. The peasants had not. The intellectual

26

16 Anatomical figure showing points on the body from which blood could be let; the illustration is taken from a fifteenth-century book belonging to the barber-surgeons of York. ▶

revolution of the seventeenth century thus did more than merely add dramatically to knowledge – it split society in a more fundamental way than the Reformation itself. In 1700 the scientists' view of the universe was totally different from that of uneducated men. That, however, is to anticipate one of the major results of the scientific revolution. It is necessary to turn now to its origins.

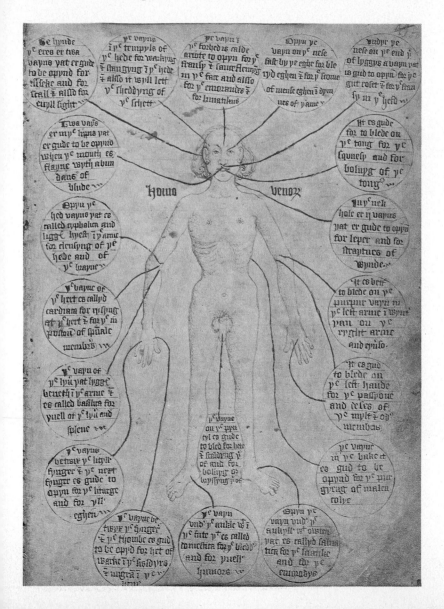

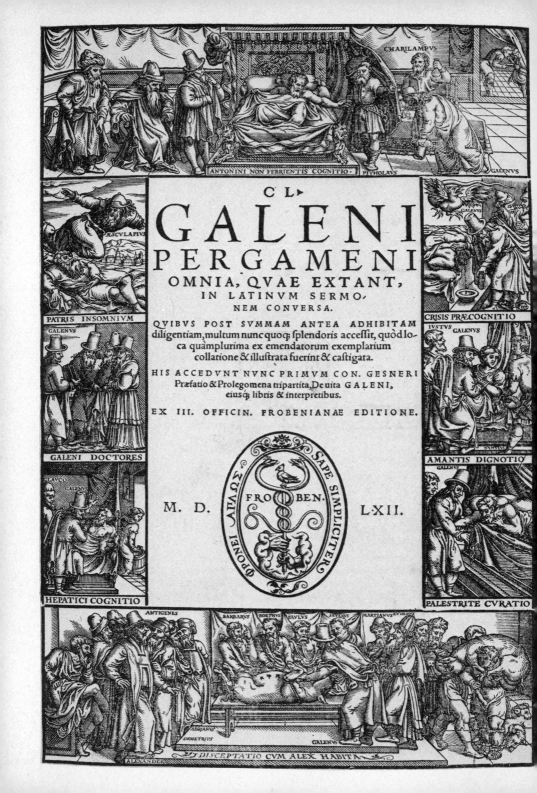

ANTONINI NON FEBRIENTIS COGNITIO. PITHOLAVS

CHARILAMPVS

GALENVS

ÆSCVLAPIVS

PATRIS INSOMNIVM

GALENVS

GALENI DOCTORES

GLAVCO GALENI

HEPATICI COGNITIO

CL.
GALENI
PERGAMENI
OMNIA, QVAE EXTANT,
IN LATINVM SERMO,
NEM CONVERSA.

QVIBVS POST SVMMAM ANTEA ADHIBITAM
diligentiam, multum nunc quoqɜ splendoris accessit, quòd loca quàmplurima ex emendatorum exemplarium collatione & illustrata fuerint & castigata.

HIS ACCEDVNT NVNC PRIMVM CON. GESNERI
Præfatio & Prolegomena tripartita, De uita GALENI,
eiusqɜ libris & interpretibus.

EX III. OFFICIN. FROBENIANAE EDITIONE.

M. D.   FRO BEN.   LXII.

ΣΩΦΡΟΝΕΙ ΣΑΦΩΣ   SÆPE SIMPLICITER

GALENVS

CRISIS PRÆCOGNITIO

IVSTVS GALENVS

AMANTIS DIGNOTIO

GALENVS

PALESTRITE CVRATIO

ANTIGENES

BARBARVS BOETHVS PAVLVS SEVERVS MARTIANVS RVDIMVS

ADRIANVS
DEMETRIVS
ALEXANDER   GALENVS

DISCEPTATIO CVM ALEX. HABITA

II THE SCIENTIFIC REVOLUTION: ORIGINS

In 1787, in the preface to the second edition of his *Critique of Pure Reason*, the German philosopher Immanuel Kant wrote that in the seventeenth century 'a new light flashed upon all students of nature' and that 'the study of nature entered on the secure methods of a science after having for many centuries done nothing but grope in the dark.' These words suggest that the cause of the scientific revolution was a burst of genius: a number of great men came to the fore who saw and abolished the mistakes of the past by the exercise of their own superlative intellectual abilities. In order to explain the origins of the scientific revolution, therefore, one need look no further than the genius of these men. Today, no historian would accept this picture without serious qualifications – far too much is now known about the medieval background to the ideas of the seventeenth-century scientists for that – but many would still argue vigorously that the genius of individuals should take the foremost place in any discussion of the causes of the scientific revolution. This type of interpretation, which stresses the role of such men as Galileo, Kepler, Vesalius and Newton, conflicts in emphasis with the two other main kinds of explanation – those which stress the medieval background, especially in mechanics and scientific methodology, to the work of the seventeenth-century scientists, and those which emphasize social and economic factors, in particular the immediate economic and religious backgrounds to the scientific developments of the period.

In an impressive discussion of the causes of the scientific revolution G. N. Clark argued that a disinterested desire to know the truth was the most important of all the motives behind it: 'truth is the same thing to the understanding as music

◀ 17 Galen's ideas on anatomy and physiology dominated medical practice for almost fifteen hundred years. The title-page shown here is taken from a collected edition of his works issued in 1561–62, nearly twenty years after Vesalius had published *De Humani Corporis Fabrica*, which marked a significant advance in the understanding of human anatomy.

to the ear and beauty to the eye', and the great scientists of the sixteenth and seventeenth centuries wanted, above all else, to satisfy their own curious minds by revealing, as far as they could, the facts about the phenomena of nature. It was, of course, mere accident that men with the intellectual energy of Galileo, Kepler and Newton were born at that particular time and devoted their powerful minds to the quest for scientific truth. Biographers of these three men and of other leading scientists of the time have described in graphic detail the stupendous mental struggles which often accompanied the birth of their

18   Large-scale mining was the source of much technological progress from the fifteenth century onwards, especially in Germany. The technical problems involved

great discoveries. Galileo's agonies as he wrestled with the profound problems of terrestrial motion, Kepler's almost inhuman persistence in the vast mathematical labours which accompanied his discovery of the three laws of planetary motions, and Newton's year and a half of concentrated mental effort when he turned his uniquely powerful mind to the task of writing the *Principia*, make it plain that the role of individual genius in the making of the scientific revolution certainly cannot be denied. If Newton had never been born the kind of synthesis which he achieved in 1687 might well have come a good deal later and

in providing pumping, hauling and ventilating machinery for the type of mine shown in this painting by Lucas Gassell (1544) stimulated research in allied fields.

in a different form, and the whole course of eighteenth- and nineteenth-century scientific developments might have been greatly altered.

This stress on the importance of genius does not fit in well with the ideas of Marxist historians and others who play down the role of the individual and consider that the economic, social and religious changes of the fifteenth to seventeenth centuries, notably the growth of capitalism, humanism and Protestantism, were the main causes of the scientific revolution.

In the view of Marxists the scientific revolution was essentially a result of the capitalist developments of the period, notably of the commercial and industrial 'revolutions' which took place in the sixteenth and seventeenth centuries and brought to the fore important technical problems which could only be adequately solved through scientific advances. Engels stated the general Marxist view very clearly in a letter of 1894. 'If society has a technical need', he wrote, 'that helps science more than ten universities. It is not that the economic condition is the cause and alone active while everything else has only a passive effect. There is, rather, interaction on the basis of economic necessity which *ultimately* always asserts itself.' Much more recently, Christopher Hill, one of the leading English historians of the early modern period, confirmed this view when he wrote that 'science . . . sprang from the shift by which urban and industrial values replaced those appropriate to a mainly agrarian society.' J. D. Bernal puts forward a similar argument in his monumental study, *Science in History*. 'It was not until the bonds of feudal order were broken by the rise of the bourgeoisie that science could advance. Capitalism and modern science were born in the same movement.' The most important factor in the appearance of the new science was 'the economic tendencies which in increasing measure throughout the later Middle Ages put a premium on technical advance, particularly in the direction of labour saving. These are the same tendencies that mark the transformation of the economic structure of feudalism into that of capitalism. Indeed, the track in time and

19 Navigating by the stars. The complex methods used by navigators and explorers stimulated scientific progress and invention.

place of the growth of capitalism in Europe is the same as that taken by the development of science.'

The plausibility of such general economic interpretations is weakened when we remember that the great scientific innovators of the period were not primarily motivated by economic forces. Copernicus and Vesalius, Galileo and Harvey did not embark upon their work because capitalist developments had made it obvious or necessary that they should do so: they produced their great discoveries because they wanted to advance the state of human knowledge and scientific truth. This, of course, is merely to say that economic factors were not basic in causing the scientific revolution, not to deny that economic conditions at the time were generally favourable to scientific advance. They certainly were favourable, and the great voyages of discovery in the fifteenth and sixteenth

20, 21   The demand for shipbuilding and navigation, born of the great voyages across the Atlantic and into the Indian Ocean (see opposite), soon produced a new class of skilled cartographers and instrument-makers. Above, a detail of Pierre Desceliers's world map, drawn in 1550, shows the North American continent. Although mariners relied on the magnetic compass to steer a course, they used astrolabes to measure celestial altitudes and so estimate their latitude. The astrolabe shown below was made in about 1588.

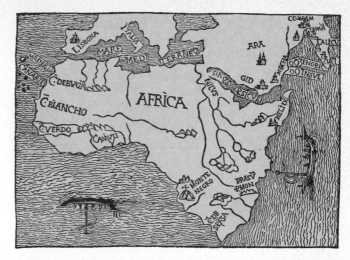

22 A remarkably accurate Portuguese map of Africa made in 1508, ten years after Vasco da Gama had sailed round the Cape to India.

23 The marine compass illustrated below was made about the same time as the astrolabe shown opposite.

centuries especially, which gave such an important stimulus to the development and spread of European trade, produced technical navigational problems which acted as a stimulus to scientific progress.

The voyages, which began with the Portuguese exploration of the west coast of Africa in the fifteenth century, extended by the sixteenth to the coastlines of Asia and of North and South America; the Portuguese pioneers were soon joined by Spaniards, Englishmen, Frenchmen and Netherlanders, all anxious to obtain their share of the prestige, wealth and territory which followed in the wake of the discoveries. The economic, political and religious motives and rivalries which led to the voyages need not be discussed here, but the explorations themselves demanded new navigational techniques and better maps, and the production of these in turn required help from learned men, notably astronomers and cartographers.

In the mid-fifteenth century, when European ships usually sailed within a limited area, the art of getting a ship from one place to another consisted almost entirely of 'pilotage' – taking ships from place to place in sight of land – as opposed to 'navigation', the art of sailing from one port to another out of sight of land. Pilotage and navigation demanded very different skills and techniques. A navigator sailing in unknown seas could obviously expect no help from the sailing directions and charts which were a pilot's stock-in-trade. His chief need was for a means of fixing the position of hitherto unknown lands so that he or his successors could find these again with the least possible trouble. The only objects by which he could determine such positions were the heavenly bodies, and the Portuguese, who pioneered the art of navigation in the course of their west African voyages, took to sea with them simplified versions of the quadrant and astrolabe – measuring devices which had long been used by astronomers ashore. It was obviously important to be able to measure as accurately as possible the latitude of a point on the earth's surface, and in 1484 King John III of Portugal appointed a commission of mathematicians to work out, with the greatest possible exactitude, a method of determining latitude at sea by observation of the sun. This is a very striking example of the way in which an urgent practical problem led through government initiative to scientific advance, for the commission produced a greatly improved version of existing mathematical tables used in fixing latitude, and also a specific procedure which enabled seamen to make use of these tables. The work of the commission was summarized in a manual for navigators which was published in Portugal under the title *Regimento do Astrolabio e do Quadrante*. This was the first European manual of navigation and it represented the best knowledge available at the beginning of the sixteenth century.

Some of the difficulties which it left unresolved, for example the problem of distinguishing the exact variation between true and magnetic north at different points on the earth's surface, were solved in the sixteenth century, but others continued to plague both seamen and governments throughout the seventeenth. The most important of these was undoubtedly that of determining longitude, and seventeenth-century governments, notably those of Charles II in England and Louis XIV in France, gave practical help and encouragement to efforts to make the measurement of longitude at sea a practicable proposition. The attempts failed. The problem of longitude is bound up with the accurate determination of time: it cannot be measured effectively without an efficient sea-going chronometer, and no such instrument was produced until the eighteenth century.

Improved techniques of navigation were closely connected with the production of accurate maps – it was little use being able to fix the position of a ship or of a newly discovered piece of land unless it was also possible to plot that position on a reliable chart for future reference. Medieval charts – 'portolans' as they were called – gave highly accurate pictures of much of the coastline of Europe, but they took no account of the fact that the world was a sphere; the area they covered was treated as a plane surface. A new form of map which would allow for the curvature of the earth and at the same time make it possible

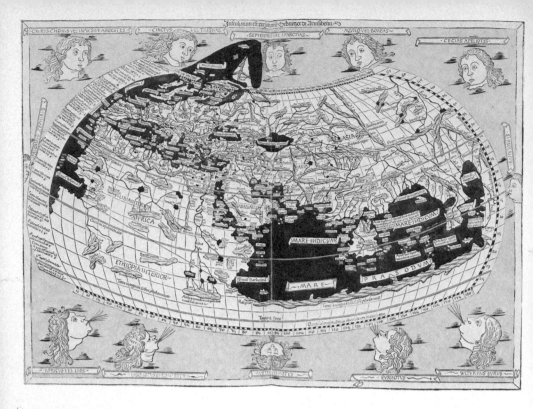

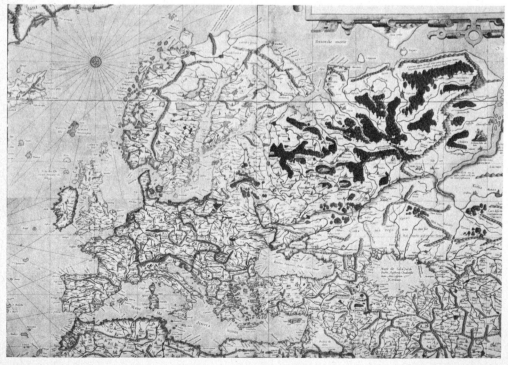

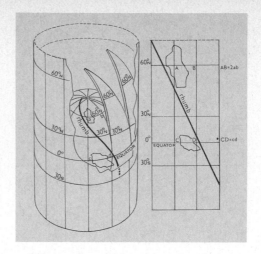

24, 25, 26 World map produced in 1482, based on Ptolemy's ideas, contrasted with a Mercator map of 1569 below, in which the course to be steered between any two points on the earth's surface lies along the straight line joining them on the map. An explanation of the Mercator principle is shown opposite, in a drawing made in 1599 by Edward Wright.

to plot a course at sea by drawing straight lines became increasingly necessary during the sixteenth century, in view of the vastly extended areas of the earth's surface which were being explored. Mercator's world map, published in 1569, met these requirements on the basis of a new projection which, in a modified form, still exists and bears his name. The mathematical theory behind Mercator's map was first explained in 1599 by Edward Wright, and charts constructed on Mercator's principles came into more and more general use during the seventeenth century, though even in 1700 many conservative navigators insisted on sticking to older forms of chart, upon which it was virtually impossible to mark discoveries with real accuracy.

There is no doubt, therefore, that the navigational problems posed by the great voyages were a significant stimulus to scientific activity. England is a good example of a country where science was closely associated with mercantile enterprise during the sixteenth and seventeenth centuries. The sixteenth-century mathematicians Robert Recorde and John Dee were both technical advisers to mercantile companies, and merchants promoted scientific education and developments through the creation of mathematical lectureships and through encouraging the translation of scientific works. In 1597 came the foundation of Gresham College in London, set up with money and property

39

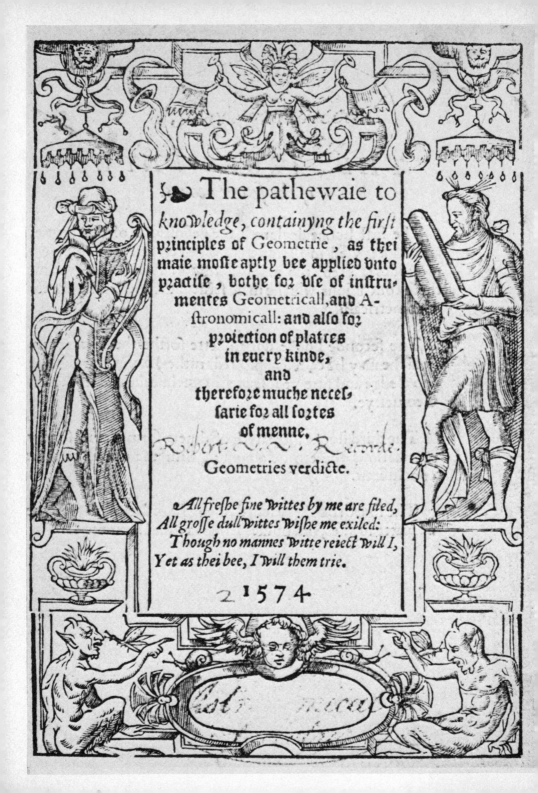

The pathewaie to knowledge, containyng the first principles of Geometrie, as thei maie moste aptly bee applied vnto practise, bothe for vse of instrumentes Geometricall, and Astronomicall: and also for proiection of plattes in euery kinde, and therefore muche necessarie for all sortes of menne.

Robert ✒ *Recorde.*

Geometries verdicte.

All freshe fine wittes by me are filed,
All grosse dull wittes wishe me exiled:
Though no mannes witte reiect will I,
Yet as thei bee, I will them trie.

2 1574

27, 28 Opposite: title-page of Robert Recorde's *The Pathewaie to knowledge*, the first work on geometry published in English (original edition, 1551). Right: John Dee, mathematician and alchemist, perhaps the greatest scientific authority of Elizabethan England.

bequeathed in the will of the great Elizabethan financier Sir Thomas Gresham, who died in 1579. Gresham had provided that three of the seven chairs in the new college should be devoted to scientific subjects and that the professor of astronomy in particular should spend a good deal of his time teaching navigational science. Gresham College became the main centre of English scientific activity in the first half of the seventeenth century, when there was a close association between the Gresham professors, the administrative officials of the English navy, and English shipbuilders and sea-captains. For example, Henry Briggs and Edmund Gunter, two of the early professors of geometry and astronomy, were friendly with John Wells, Keeper of the Royal Naval Stores at Dartford, and Gunter, Wells and a number of English naval architects succeeded in working out a more accurate method of calculating the tonnage of ships.

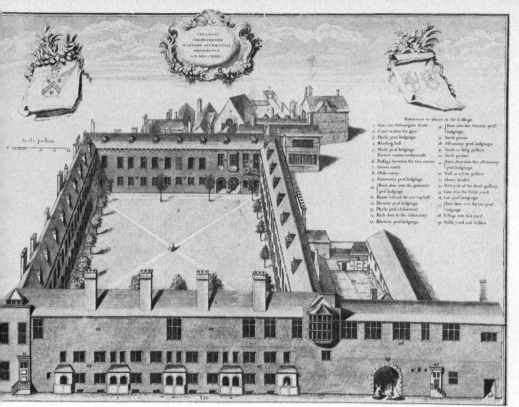

30   An engraving of Gresham College, made in 1739 by George Vertue.

There were other forces in the society of the time, besides the great voyages, which played a part in the advance of science. One of these was the agitation for calendar reform, which had strong religious overtones. The cumulative errors of the existing Julian calendar had long been recognized, and proposals for reform dated from the thirteenth century or even earlier. These remained ineffective for many years, but by the sixteenth century the increasing complexity of economic and social life put an obvious premium upon an efficient method of dating, and reform received enthusiastic backing from the Church, with results for astronomy which can be illustrated from the life of Copernicus himself. Copernicus was asked

43

◀ 29   Sir Thomas Gresham, builder of the first Royal Exchange in London and founder of Gresham College, portrayed by Antonio Mor. He is still remembered for Gresham's Law – bad money will drive out good if both are in circulation together.

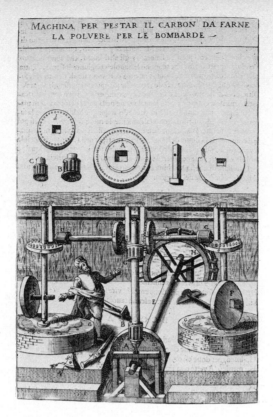

31 By the early seventeenth century, many gunpowder mills were using water-power to drive their heavier machinery.

early in his career to advise the Papacy on calendar reform, but he declined on the grounds that existing astronomical observations and ideas did not provide an adequate framework for the construction of an accurate new calendar. In other words, Copernicus believed that reform of the calendar had to be preceded by the modification of astronomical beliefs, and one, though not the most important, of the reasons which led him to consider his heliocentric theory of the universe was, as he himself said, the fact that 'the mathematicians are so unsure of the movements of the sun and moon that they cannot even explain or observe the constant length of the seasonal year.' He hoped that his own work, published in 1543, might make a new calendar possible and the Gregorian calendar, which was adopted by the Papacy in 1582, was based to some extent upon his computations.

44

Other important social forces which stimulated scientific advance were connected with medicine and warfare. In all ages there is a desire to improve the lot of man through greater understanding of his body and its functions, and the advances in anatomy, physiology and the treatment of disease which took place in the sixteenth and seventeenth centuries as a result of the work of such men as Paracelsus, Vesalius and Harvey were motivated in part by a genuine desire to improve man's individual lot and the general social fabric by the conquest or amelioration of bodily ills. In addition states required more efficient weapons for use against their rivals, and this led to efforts to harness the work of scientists to the needs of warfare. Governments were very anxious, for example, to obtain the most effective possible methods for computing the range of cannon and this may have given a good deal of impetus to work on dynamics, including that of Galileo himself.

Humanism was another force in early modern society which played a part in bringing about the scientific revolution. The general revival of interest in classical antiquity which was such a marked characteristic of the work of fifteenth- and sixteenth-century humanists extended to the writings of the great scientists of the Greek and Roman periods, including those of Archimedes and Galen. The virtual rediscovery of Archimedes' works in the early sixteenth century gave a great stimulus to mechanics

32 Angling a cannon: woodcut from a sixteenth-century manual on civil and military engineering.

45

and mathematics, and improved editions and translations of already familiar works of Galen, as well as the discovery and publication in 1531 of a hitherto unknown Galenic text, *On Anatomical Procedures*, gave a great boost to the study of anatomy. Vesalius's *De Fabrica*, published in 1543, which superseded Galen in the field of anatomy and was one of the seminal works of the scientific revolution, was itself founded to some extent on the *Anatomical Procedures* and on another of Galen's works, *On the Use of the Parts*, which was also virtually unknown before Vesalius's lifetime.

The rediscovery of classical scientific texts must, therefore, have a place in any discussion of the origins of the scientific revolution, but it is, of course, a limited place. The humanists of the fifteenth and sixteenth centuries looked back to classical antiquity as the golden age of the world, a period which could hardly be equalled, let alone surpassed, by contemporaries, and this idea extended to their view of the science of the ancients. A great feature of the scientific revolution, however, was that it did improve very considerably upon the scientific ideas of the ancient world. The rediscovery of classical texts, which gave a significant impetus to scientific studies, cannot, therefore, account for the direction which these studies finally took.

Historians who emphasize the significance of contemporary conditions in European society in explaining the scientific revolution often ascribe an important role to the Reformation and the development of Protestantism, particularly in its Calvinist varieties. This idea can be found in the work of such distinguished scholars as Weber and Hill. Their argument, put in the most general terms, is that Protestantism contributed to the rise of modern science by replacing the hierarchical authority of the Catholic Church, with its emphasis on the power of the ordained priesthood, by the less rigid organizations of Protestantism, which stressed the priesthood of all believers and exalted the rights of the individual conscience. The logic of Protestantism, even though this was not immediately recognized by the leaders of the Protestant churches,

33 Improved translations of Galen's writings played an important role in advances in anatomical knowledge: here Galen is portrayed in a sixteenth-century woodcut.

with their overwhelming belief in a literal interpretation of the Bible, was, therefore, to promote a spirit of free inquiry in all fields, and, of course, modern science and the spirit of free inquiry go hand in hand.

This interpretation has recently been applied in detail to English conditions by Christopher Hill in his book *Intellectual Origins of the English Revolution*, where he argues that Puritanism and modern scientific ideas developed together in opposition to the crypto-Catholicism and scientific obscurantism of early seventeenth-century Anglicanism and its royalist supporters. Hill makes this point in various ways: by reference to the Puritanism of some of the early Gresham professors, such as Henry Gellibrand and Samuel Foster; by associating Puritanism and freedom of thought; by arguing that the new science and Puritanism triumphed together during the 1640s and 1650s; and by emphasizing the links between the Puritan demands for first-hand religious experience and Francis Bacon's insistence on a science based on personal observation and experiment. Hill's ideas have attracted a good deal of attention among historians, much of it critical. It has been objected, for example, that he stresses those factors of Puritanism which encouraged men to turn to science, but neglects the anti-scientific trends – notably otherworldliness and anti-intellectualism; these were

47

undoubtedly important features of Puritanism and have been emphasized in all important recent considerations of the movement. It is also difficult to distinguish many radical Protestants among the prominent English scientists of the time. It is true that there were a few – Henry Briggs, Jeremiah Horrocks and John Napier are obvious examples – but most leading English scientists, including men as distinguished as William Harvey and Robert Boyle, seem to have belonged to a quite different tradition of religious thought which can best be described as moderate or latitudinarian. Moreover, Francis Bacon, who stands at the centre of many of Hill's most important arguments, was not himself a Puritan, and the attempt to set his ideas in a Puritan context is not always convincing.

If Hill's attempt to demonstrate an intimate link between religious radicalism and the development of English science has not met with general acceptance, the same can be said for the ideas of those who have argued, on a European basis, that the rise of Protestantism was one of the major causes of the scientific revolution. It has been pointed out, for example, that there is no reliable statistical evidence to support this view. Alphonse de Candolle's *Histoire des Sciences et des Savants depuis deux Siècles*, written in the later years of the nineteenth century, which shows that Protestants outnumbered Catholics among European scientists from the 1660s onwards, does not provide evidence for the years before 1660, the crucial period in the genesis of the scientific revolution. It is true that statistics have been produced to show that there were more Protestants than Catholics among the scientists of the southern Netherlands in the later years of the sixteenth century, but there are doubts about the reliability of these figures, and, even if they are accepted, it is difficult to deny that Catholic Italy enjoyed a scientific predominance throughout that century which puts the contributions of Protestant countries into a secondary place.

Throughout the sixteenth century, in fact, Protestant religious leaders were far more conservative in their attitude towards new scientific theories than their Catholic contemporaries. Luther, Melanchthon and Calvin all hastened to condemn Copernicus. In striking contrast, for over seventy years after Copernicus's death the official voice of Catholicism remained silent about his theories, which were not condemned by the Church until 1616. It seems to have been this condemnation and the trial of Galileo which followed seventeen years later which caused a drastic decline in Catholic scientific inquiry. By the middle of the seventeenth century the forces prevalent in Counter-Reformation Catholic society were inimical to scientific advance, just at the very time when Protestant theologians were becoming much more flexible in their attitude towards the new discoveries. Thus Protestant and Catholic attitudes towards the new science changed together, but not before Catholic scientists had made a decisive contribution to the great discoveries of the pre-1660 period. The predominance of Protestants among scientists in the post-1660 years reflects this mid-century change of attitude.

The third major kind of explanation of the scientific revolution, the one which emphasizes its medieval origins, has emerged only during the last half-century as a result of the work of a number of distinguished medieval historians headed by the great French scholar Pierre Duhem. They have shown that both the methodology and the content of the work of Galileo and his contemporaries were anticipated in important ways in the thirteenth and fourteenth centuries by scholars at the universities of Oxford and Paris.

In the field of methodology the decisive breakthrough seems to have come in thirteenth-century Oxford, where important advances were made on Greek scientific method. The Greeks tried to express much of their science in the form of logical deductions from indemonstrable but unchallengeable first principles. This type of approach, which was described by Plato in the *Republic*, was shared by Aristotle. It is true that, unlike Plato and his followers, Aristotle and his supporters did seriously discuss the inductive method – the building of theories on observed facts – but this was not central to the Aristotelian idea

49

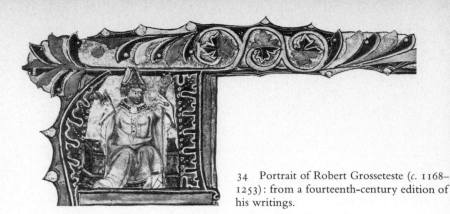

34 Portrait of Robert Grosseteste (*c.* 1168–1253): from a fourteenth-century edition of his writings.

of science, which was based essentially on definitions of the 'nature' or 'essence' of a thing. For example, Aristotle believed it was the 'nature' of heavenly bodies to move in circles and of terrestrial bodies to move either upwards or downwards in straight lines. That, according to him, was a full and sufficient 'explanation' and from it he proceeded to *deduce* important consequences. Such ideas, which at best relegated induction to a secondary role in scientific explanation, could never have provided a satisfactory basis for the scientific revolution, which required a very different attitude towards induction and experiment for its achievements. Two fundamental methodological problems were involved. The first was how to investigate as comprehensively as possible a phenomenon or series of phenomena and then to generalize the observations obtained into theories which might explain them: that was the problem of induction. The second problem was how to distinguish true and false among such theories by further investigations: that was the problem of experimental verification.

The first clear understanding of these problems appears in the thirteenth century in the writings of the great English scholar Robert Grosseteste, and under his influence and that of his followers the new methodology made rapid progress at Oxford, where it was used by Grosseteste in a detailed study of optics. Grosseteste and his school also appreciated the significance of mathematics – another of the tools which was fundamental in bringing about the seventeenth-century scientific revolution. Roger Bacon, an important thirteenth-century Oxford scientist, who was one of the most distinguished of Grosseteste's followers, made the point well when he wrote: 'All categories depend on a knowledge of quantity, concerning which mathematics treats, and therefore the whole power of logic depends on mathematics.' And again, 'In the things of this world, as regards their efficient and generating causes, nothing can be known without the power of geometry.' The thirteenth-century Oxford scholars, however, despite their realization of the importance of mathematics, did not attempt precise calculations in their experiments. This was no doubt partly due to the primitive mathematical concepts of the time, but it also seems to have been due to a kind of 'stop in the mind': they were uncertain how to proceed in practice.

35 Links between English and other European scientists in the Middle Ages were close. This picture shows Roger Bacon presenting a book to the Chancellor of Paris University.

These ideas about scientific methodology soon spread through the continent of Europe, to France, Germany and Italy. They were widely current by the sixteenth century and formed a significant part of the mental background of Galileo. However, though they marked a vital stage on the way to the scientific revolution, these ideas needed extra ingredients, which were provided by Galileo in the seventeenth century, before they could emerge in a recognizably modern form. Thirteenth- and fourteenth-century scientists were concerned with the new methodology mainly for its own sake; they were not really interested in *applying* it in a systematic way to specific occurrences in nature. Galileo, on the other hand, experimented on actual natural phenomena and also made specific measurements which enabled him to express his results in quantitative form.

In the content as well as in the methodology of science, medieval developments were important in the genesis of the scientific revolution. The revolution in astronomy was really initiated by the publication of Copernicus's heliocentric ideas in his great work of 1543, but he was not the first man in Europe to raise the question of the earth's motion. Eminent thinkers of antiquity had challenged Aristotle's doctrine of a motionless earth and in the fourteenth century a French scholar, Nicole Oresme, who became Bishop of Lisieux, discussed at length the possibility that the earth rotated daily on its own axis, showing that this theory could not be dismissed either on grounds of argument or observation. A hundred years later Nicholas of Cusa, a cardinal and a distinguished philosopher, also discussed the motion of the earth. Neither Oresme nor Nicholas of Cusa was a professional astronomer and neither constructed a detailed system of the universe which could challenge the cosmology of Aristotle or the astronomy of Ptolemy, but their work does show that the fundamental Aristotelian concept of a motionless earth was questioned in the Middle Ages, long before Copernicus's time; and Copernicus was certainly acquainted with Cusa's teachings.

36 Nicole Oresme with his armillary sphere, an astronomical device with a number of rings representing the equator, tropics, the sun's apparent orbit, and so on. It was used to determine the position of the stars.

It was, however, in the study of motion that the Middle Ages made what was probably its most important contribution to the scientific revolution. In the fourteenth century a satisfactory law of acceleration was established for the first time, and important advances were made in theoretical discussions about freely falling bodies and the motion of projectiles. Correct ideas about acceleration were an essential part of seventeenth-century science – the great work of Galileo and Newton would have been impossible without them – but Aristotle's views on the subject were fragmentary and misleading, and it was only in the first half of the fourteenth century that a satisfactory treatment of the problem emerged at Merton College, Oxford, as a result of the work of a small band of scholars.

In about 1335 one of these men, William Heytesbury, gave the first correct statement of a law of uniform acceleration, which can be paraphrased as follows: 'A moving body either accelerating or decelerating uniformly during a given period of time covers exactly the same distance as it would have covered in the same period of time if it had moved steadily at the speed it would have achieved at the middle moment in time of its uniform acceleration or deceleration.' This can be illustrated by imagining a car accelerating uniformly from rest to a speed

53

of 60 miles per hour over a distance of 2 miles, and another car moving uniformly at 30 m.p.h. over the same 2 miles. Both cars would complete the trip in exactly the same time, but halfway in time through the journey, the car moving uniformly at 30 m.p.h. would have covered exactly half the distance, whereas the car accelerating from rest, which would also by then have achieved a speed of 30 m.p.h., would have covered only a quarter of the distance. It would cover the remaining three-quarters while accelerating uniformly from 30 to 60 m.p.h. Time was therefore the crux of the matter and the Merton scholars were able to draw the correct conclusion that a body accelerating uniformly from rest would cover three times as much ground in the second half of its travel as it covered during the first half of the time.

This acceleration law was proved in a number of ways during the fourteenth and fifteenth centuries, the most important being that employed by Nicole Oresme. Oresme used graphs in his proof, and it has been argued that his graphing system anticipated, in some important ways, the calculus, which only emerged in a mature form in the seventeenth century as the most important mathematical tool of the age of Newton.

The fourteenth-century scholars of Oxford and Paris, who thus propounded and proved a satisfactory law of acceleration, do not seem to have applied it to freely falling bodies. It was left to Galileo to do this in some of his most famous experiments, and the contrast between the 'theoretical' approach of the Schoolmen on the one hand and the 'practical' experiments of Galileo on the other is yet another reminder of the differences between fourteenth- and seventeenth-century scientific attitudes, differences which help to explain why the vision and achievements of the Oxford and Parisian scholars of the Middle Ages were so much more limited than those of Galileo.

Despite this failure to apply Heytesbury's acceleration theorem to objects in free fall, there were important discussions in the fourteenth century which carried ideas about such bodies beyond those of Aristotle. Aristotle and his followers had

54

37   Page from a fifteenth-century copy of Nicole Oresme's treatise *De Configurationibus Qualitatem*, which proves the theory of uniform acceleration. ▶

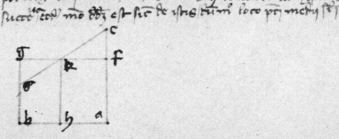

explained the acceleration of falling bodies either in terms of a naturally increasing speed the nearer they drew to their home in the universe or else as a result of the increased pressure of the air above them as they fell towards the earth. These, together with other ideas of Aristotle's about motion, had been criticized by John Philoponos, a Byzantine philosopher of the sixth century, who had introduced the idea of an incorporeal moving power which could be incorporated into a body by the original mover or movement and thus keep the body in motion. This idea was adopted and developed by Jean Buridan, a notable French scholar, who was Oresme's teacher. Buridan thought of the incorporeal power as a kind of inner energy in the body affected. He called it impetus, and applied his ideas to both projectiles and bodies in free fall. He set out his thoughts on the acceleration of falling bodies in one of his commentaries on the works of Aristotle.

> A heavy body not only acquires motion unto itself from its principal mover, i.e. its gravity, but it also acquires unto itself a certain impetus with that motion. This impetus has the power of moving the heavy body in conjunction with the permanent natural gravity. . . . From the beginning the heavy body is moved by its natural gravity only; hence it is moved slowly. Afterwards, it is moved by that same gravity and by the impetus acquired at the same time; consequently it is moved more swiftly. And because the movement becomes swifter, therefore the impetus also becomes greater and stronger, and thus the heavy body is moved by its natural gravity and by that greater impetus simultaneously, and so it again will be moved faster; and thus it will always and continually be accelerated to the end.

Buridan was saying in effect that a primary force, gravity, was impressing a secondary force, continually increasing impetus, into the falling body, which thus accelerated continuously. These ideas were clearly much more sophisticated than those of Aristotle and they represented a real advance in understanding. It was to them, and to other medieval ideas about the movement of projectiles, that Galileo turned when he embarked on his own studies of motion. In due course, in his experiments with freely falling bodies, he came to reject the idea of impetus as an intermediate force caused by gravity and itself *causing* acceleration, and saw it instead as merely a *measure* and *effect* of the force of gravity. This led naturally to the classical position, enshrined in Newton's second law of motion, that a continually acting force, gravity for example, results in acceleration.

Buridan also made notable contributions to the theory of projectiles. Aristotle argued that the force of air rushing round behind a projectile caused its continued movement after contact with the original motor ceased. The scholars of fourteenth-century Paris, in contrast, applied their theory of impetus to the study of projectiles; but the differences between two views of impetus current among them were of considerable importance in their ideas about projectile motion. One view, which was held by Buridan's pupil Oresme, was that the force of impetus applied to a projectile tended to die away naturally, and eventually disappeared altogether of its own accord. The other, expounded by Buridan, was that the force of impetus was capable of lasting indefinitely as long as nothing interfered with it. As he himself put it: 'Impetus is a thing of permanent nature. . . . It is a quality naturally present and predisposed for moving a body in which it is impressed. . . . It is impressed in the moving body, along with the motion, by the motor; so with the motion it is remitted, corrupted or impeded by resistance or a contrary inclination.' And elsewhere he wrote, 'The impetus would last indefinitely if it were not diminished by a contrary resistance or by an inclination to a contrary motion.'

Galileo took up, modified and developed medieval ideas about projectiles. He came to accept Buridan's idea of impetus as permanent rather than self-expending, but, as already noted, also ended up by regarding it as a *measure* rather than a *cause* of motion. In doing this he discovered the law of inertia, which

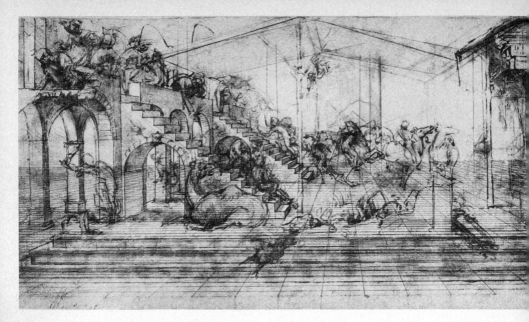

38 The development of perspective: study for Leonardo da Vinci's *The Adoration of the Magi*.

is the basis of modern theories of motion, and, as enshrined in Descartes's famous definition and Newton's first law, forms one of the most important achievements of the scientific revolution.

These developments in mechanics were an immensely important contribution by the Middle Ages to the scientific revolution, but other intellectual trends and practical advances which can be traced back to the fifteenth century or earlier also had a considerable influence on the work of the sixteenth- and seventeenth-century scientists. Three of the most important of these were developments in the theory of art, the Platonic revival of the later Middle Ages, and the invention of printing in Europe.

Much of medieval painting may now seem lifeless and unreal, and, although conscious efforts in the direction of naturalism were made by Giotto at the beginning of the fourteenth century, it was not until the fifteenth that artists finally realized that the key to naturalistic representation of scenes and people lay in a scientific study, founded upon geometry, of the problems of perspective. By 1425 Brunelleschi had worked out a satisfactory system of perspective, and he spread his ideas through pupils such as Donatello and Masaccio. The first written account of the new methods, Alberti's *Della Pittura*, was published in 1435. Underlying all Alberti's theories was the idea that painting was concerned with the accurate representation of the visible world, and he stressed the importance of mathematics for the artist. 'No painter', he maintained, 'can paint well without a thorough knowledge of geometry.' Renaissance artists, indeed, were often practising mathematicians and were frequently called upon to do service as architects, engineers and advisers on ballistics. Artists who were also respected scientists were therefore familiar figures in the fifteenth century, and this tendency to link art and science culminated in the early sixteenth century in the work of that universal genius, Leonardo da Vinci, whose magnificent artistic achievements were linked with his important scientific studies of the anatomy of the human body.

All the characteristics of the work of the fifteenth-century artists – accurate representation of man and of the real world, attention to details and dependence on mathematics – were also important features of the scientific revolution. To take just one example, Vesalius, who inaugurated the revolution in anatomy, could never have produced his *De Fabrica* without the aid of Renaissance artistic achievement. Its illustrations of the human body are a monument to naturalism and a tribute to the advances which the artists of the fifteenth century had made possible.

The interest in mathematics which was such a feature of the sixteenth and seventeenth centuries had one of its roots in the mathematical enthusiasm of the Renaissance artists; but another and probably much more important root can be found in the revival of Platonism in the later Middle Ages. The ideas of Plato, with their emphasis on the importance of mathematics in explaining the universe, had been very influential in early

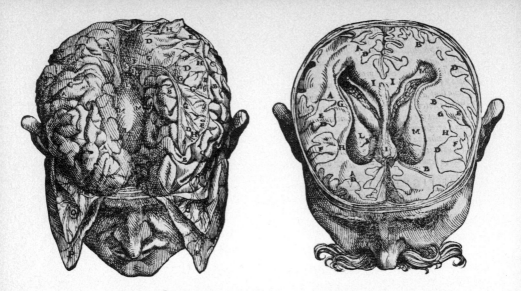

39, 40, 41 Above: diagrams from Vesalius's *De Fabrica*, showing dissections of the human brain. Opposite: Leonardo da Vinci's drawings of the human foetus, a perfect example of artistic achievement and scientific investigation combining to advance knowledge.

medieval Europe, and even in the thirteenth century, when the views of Aristotle captured medieval thought, Platonic ideas were by no means routed and can be traced in the work of such important thinkers as Roger Bacon in the thirteenth century and Nicholas of Cusa in the fifteenth. Nicholas believed that the theory of numbers was the essential element in the philosophy of Plato and that the universe was a perfect harmony in which all things had their mathematical proportions. There was indeed a strong revival of Platonism in southern Europe in the fifteenth and sixteenth centuries, and this development, which challenged the dominant Aristotelianism with its view of the low importance of mathematics, had a powerful influence on such seminal figures of the scientific revolution as Copernicus and Kepler. In his youth Copernicus spent six years in Italy, and during his studies at the university of Bologna became friendly with one of his teachers there, Dominicus de Novara, who was professor of mathematics and astronomy. De Novara was an enthusiastic protagonist of Platonic ideas

60

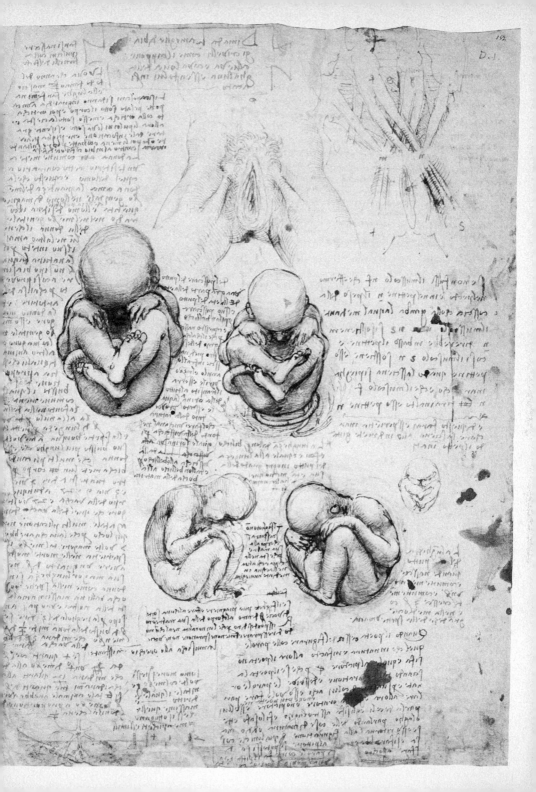

and was a leading critic of the Ptolemaic system in astronomy, which he thought both cumbersome and disorderly and a gross violation of the Platonic ideal of the universe as a harmonious mathematical system. Copernicus imbibed some of his master's ideas, and his subsequent attack on the Ptolemaic system, which was largely designed to rationalize and simplify it, can be seen, in part at least, as an application of Platonic ideas to the astronomical problems of the day.

Another of the great minds of the scientific revolution who was strongly influenced by Platonism was Johannes Kepler, who wrote: 'Just as the eye was made to see colours, and the ear to hear sounds, so the human mind was made to understand, not whatever you please, but quantity.' Quantity was the basic feature of things, the *primarium accidens substantiae*. Kepler believed that the real world should be understood in terms of mathematical harmonies. It was during this vain quest that he discovered the three laws of planetary motions, which are the foundation of his later fame as one of the giants of the scientific revolution.

The ideas of Kepler and his fellow scientists could, of course, be circulated in printed books, and this is a fact of immense importance. Medieval scholars could only distribute their works in manuscript copies, a slow and uncertain process. There was no guarantee that important scientific work done in one part of medieval Europe would be readily available in other parts of the Continent. Printing changed all that. Its invention in Europe in the mid-fifteenth century – an invention independent of the long-established printing techniques of the Chinese – meant that scientific books, like all scholarly works, were quickly made available throughout Europe. Printing-presses spread with great rapidity. By 1500 they were to be found in every European country except Russia – over 1700 in almost 300 towns. It has been estimated that almost 40,000 editions of books of all kinds were published during the fifteenth century and the expansion of the industry continued during the sixteenth and seventeenth centuries.

THEODOSII

DE SPHAERICIS LIBRI TRES, A IOANNE VOGELIN
Hailpronnenſi, Aſtronomiæ, in Viennenſi Gymnaſio, ordi-
nario profeſſore, Ciuilisq̃ collegij, collega, reſtituti,
& Scholijs non improbandis illuſtrati.

Cum gratia & Priuilegio Regiæ Maieſtatis.

Der Swangern frawen vnd
Hebammen Rosegarten

42, 43, 44   These two title-pages, from a book on cosmology (left) and from the
first printed work on midwifery, demonstrate the high standard of sixteenth-
century printing. Below: a contemporary printing shop, showing the various stages
from composition (left) to printing the sheets (right).

Numerous scientific works were included among the flood of books. They were read by three main classes of people. First of all, there were the professional scholars, who were able to study the ideas of their colleagues and rivals very soon after these had been formulated; secondly, men to whom scientific developments were of practical concern – for example, the literate sailors who were eager customers for manuals of navigation; and finally, there was the educated élite among the general public, who, in the seventeenth century, absorbed the broad general ideas of the scientific revolution from the books which popularized them.

The scientific revolution, which could hardly have taken place without the invention of printing, did indeed owe a debt to most of the influences which we have been considering, from medieval advances in philosophy, artistic theory, mechanics and scientific methodology, to early modern economic and social developments and forces, such as the voyages of discovery and problems of warfare and medicine.

When all is said and done, however, the scientific revolution was the work of a handful of great scientists. However much they owed to a medieval tradition or to a favourable economic and social background, it was these men with their seminal minds who actually made the fundamental discoveries and experiments which constituted the vital changes of the period. In the last resort the scientific revolution was accomplished by the genius of such men as Copernicus, Tycho Brahe, Kepler, Galileo, Descartes and Newton in astronomy, mechanics and mathematics; of Vesalius and Harvey in medicine; and of Bacon, Galileo and Descartes in methodology.

45   An early seventeenth-century paper-mill driven by an undershot water-wheel. The large-scale manufacture of paper, together with the invention of printing from movable type, made possible the rapid dissemination of information to an ever-growing number of readers. ▶

46   Sir Isaac Newton: portrait by Sir Godfrey Kneller.

III THE SCIENTIFIC REVOLUTION: THE GREAT DISCOVERIES

THE NEW MATHEMATICS

The work of the great seventeenth-century scientists depended on spectacular advances which took place at that time in mathematics, in scientific methodology and in the making of scientific instruments. These developments, which the scientists themselves largely brought about, made possible the dramatic discoveries of the period in astronomy and mechanics, discoveries which were at the centre of the scientific revolution. It is important to notice too, that, at the end of the seventeenth century, the work of scientists, who were already by then conducting an elaborate correspondence among themselves, began to be systematically pooled and the results made widely available through the creation of new learned societies and the publication of journals which were sometimes specifically associated with these societies.

Although important mathematical developments, especially those associated with the name of Vieta (the Frenchman François Viète, who died in 1603), took place in the sixteenth century, they were far surpassed in importance by the revolutionary innovations of the next century, one of the greatest periods in the whole history of mathematical progress. It is difficult to say precisely why there should have been such spectacular mathematical achievements at this particular time, but the main reason was probably the emergence of a number of mathematicians of genius, each of whom made new discoveries and at the same time built on the work of his predecessors. There were improvements in virtually all branches of the subject. Most of the symbols now regarded as essential in mathematical operations were either invented or came into general use in these two centuries: this was true of the signs for

47, 48   Sir John Napier (1550–1617), whose invention of logarithms was first made public in 1614 in his *Descriptio*, two pages of which appear opposite.

addition (+), subtraction (−), multiplication (×), division (÷), equality (=), greater than (>), and less than (<). The seventeenth century also saw the introduction of brackets and of the decimal system. The importance of the use of symbols was very clear in Vieta's work. His main contribution was in the field of algebra, where he rationalized the symbols used and reduced their number, making algebra for the first time a purely symbolic science and raising it to a much higher level of generality and abstraction than ever before. This was a most significant step in the history of mathematics, which depends for much of its progress on a continual generalization of more specialized ideas and concepts. Vieta's work was an important prelude to the great discoveries of his seventeenth-century successors, who surpassed him in their achievements but whose fundamental mathematical attitudes were much the same as his own.

A further significant advance in mathematical technique, the invention of logarithms, was made in the early seventeenth century by the Scotsman John Napier, who published his discovery in his *Descriptio* of 1614. Logarithms reduced the operations of multiplication and division to addition and subtraction, thus saving an immense amount of labour in calculations, especially where very large numbers were involved. The value of Napier's work was immediately recognized by his friend Henry Briggs, professor of astronomy at Gresham College, who in 1624 published his *Arithmetica Logarithmica*, which gave the common logs of thirty numbers to fourteen decimal places. Indeed, Napier's invention was enthusiastically taken up throughout much of Europe, and logarithms were soon being used in Italy, Germany and France. They were particularly useful in the increasingly elaborate astronomical calculations of the day, and the great French astronomer Pierre de Laplace later asserted that their invention, 'by shortening the labours, doubled the life of the astronomer'.

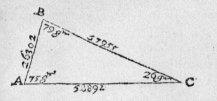

Logarithme of 75 degrees, and a little more wch is the angle A sought for: if A appeare to be an acute angle, otherwise 105 deg (by the 1 and 2 Sect. chap. 3. book. 1.) if it appeare to be an obtuse angle.

Contrariwise, if the angle A bee giuen 75 degr. and the angle C, and the side B C as before, and A B be sought for.

Adde the Logarithme of B C + 545471—0
to the Logar. of the angle C + 824689

they will be made as afore ——+1370160—0
From which take the } —+ 34667 ;
Logar. of the angle A }

There will come forth   + 1335492—0 the Logarithme of the side A B, and the number thereof 26302, which was sought for.

The angles A 75 degr. and C 26 deg. being now found, the angle B shal be 79 deg (by the 3. *Prop.* of this book:) out of which being now found, the side opposite thereto A C 58892 is no otherwise found then the side opposite thereto (A B) was lately found by the angle C. Therfore now all the parts of this oblique-angled triangle are knowne.

In the obliquangled triangles, we call them legs which

which are about any angle, & the base which subtendeth the same.

In obliquangled triangles, the Logarithme of the *Propos.5.* summe of the legges, subtracted from the summe made of the Logarithme of the difference of the legs, and the Differentiall of halfe the summe of his opposite angles, leaueth the Differentiall of halfe the difference of the same.

Because as the summe of the legges is to the difference of the legges; so is the Tangent of halfe the summe of their opposite angles to the Tangent of halfe the difference of the same: Therfore they are proportionall, and by the 1 *Prop.* 2. *Chap.* 1 *Book.* the differences, or excesses of their Logarithmes are equall. Therfore (by the 4. *Prop.* 2. *chap.* 1. *book*) we must necessarily conclude as before.

Therefore by two legs, and the angle contai- *A Corolarie* ned betweene them, are knowne by this Proposition, the other opposite angles, and thereby the other side, by the proposition going before.

For the Logarithme of the summe of the leggs being subducted out of the summe made of the Logarithme of the difference of the leggs, and the Differentiall of halfe the summe of the opposite angles put together, there shall come forth the Differentiall of halfe the difference of the same angles; which halfe difference being added to the halfe summe aforesaid, there shall come forth the greater angle; and being subtracted, the lesse.

As in the foresaid *Obliquangled triangle* A B C
Let there be giuen A B one legg   26302
           B C th'other leg   57955
           B the angle contained
betweene them, 79 degrees, and let the other angles

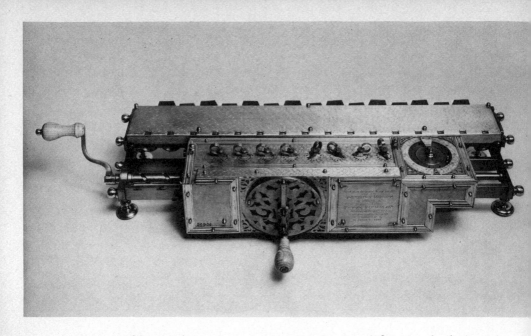

If logarithms were important in simplifying calculations in astronomy, the analytical geometry associated with the name of René Descartes was immediately useful in the solution of problems in mechanics, that other vital area in contemporary scientific developments. Descartes's *Géometrie*, which was published in 1637, annexed to his *Discourse on Method*, was the first treatise on analytical geometry. In it he showed that geometric problems could be put into algebraic forms and could be solved by numerical calculations based on algebraic and arithmetical methods. This technique was of great use in the dynamical problems associated with ballistics, where it helped in working out ranges, as well as in numerous other calculations in the science of mechanics.

By far the most significant mathematical development of the period was, however, the invention of the calculus. This was essentially the work of two men of genius, Isaac Newton and Gottfried Leibniz, who synthesized and completed the ideas of other men. There has been considerable dispute among historians of science as to which of the two should be given the main credit for the invention. It now seems clear, however,

, 50 Portrait of Gottfried Leibniz
646–1716). His calculating machine,
posite, was used for multiplying,
viding and extracting roots, as well
for adding and subtracting.

that each came to his conclusions quite independently of the
other. Newton made his discoveries first, in the 1660s, but did
not publish them until 1704. Leibniz invented his form of the
calculus in the 1670s, but his results were published in 1696 in a
textbook by Guillaume de L'Hôpital. Each man, therefore,
deserves credit for the creation of one of the most important
mathematical tools of all time, but it is a pity that Leibniz's
unquestionably superior notation was not adopted in the
English-speaking countries of the world until the nineteenth
century. The significance of the calculus was that it made possible the sophisticated mathematical investigation of continuous change in all its forms, both in pure mathematics and
in the sciences. Its importance in mechanics and astronomy,
where problems connected with the changing movement of
bodies are paramount, is too obvious to need stressing, and its
general importance in the history of mathematics was well
summarized by E. T. Bell when he wrote that with its invention
'creative mathematics quite generally passed to an advanced
level and the history of elementary mathematics essentially
terminates.'

The mathematical developments which we have just examined helped to bring about some of the great scientific discoveries of the period – for example, Newton used his calculus in the composition of the *Principia* – but the new scientific method which was created in the seventeenth century was even more fundamental for the achievement of the scientific revolution. Bacon, Descartes and Galileo, though its most important creators, owed a debt to those medieval philosophers who had shown an appreciation of the problems of induction and experimental verification and a knowledge of the importance of mathematics in their scientific work. These medieval advances, which were part of the general intellectual inheritance of the early seventeenth century, required elaboration and development in significant ways before a 'modern' scientific method could be produced.

The medieval philosophers had not really experimented on nature in a systematic way – they were more interested in the *theory* of their new methodology than in its application to the real world. By contrast, Francis Bacon emphasized the importance of experiments and laid great stress on their *systematic* use to build up a body of empirical knowledge from which general theories could be established and tested. Existing interpretations of nature, he maintained, were usually 'founded on too narrow a basis of experiment'. What was needed was a host of new experiments in all the fields of science. These should be properly organized and recorded, and experimenters in different fields should co-operate in order to achieve a fruitful exchange of ideas which might lead to further experiments and further advances. Once a sufficient number of phenomena had been investigated, general theories could be produced by induction, the method of reasoning which, as he described it in his *Novum Organum*, published in 1620, 'derives axioms from ... particulars, rising by a gradual and unbroken ascent, so that it arrives at the most general axioms last of all. This is the true way but as yet untried.' These axioms, once they had been established, could

72

51   Title-page of *Instauratio Magna*, the first part of Francis Bacon's unfinished work on the reorganization of the sciences. The book, which appeared in 1620, contains the *Novum Organum*, in which Bacon sets out his views on scientific method. ▶

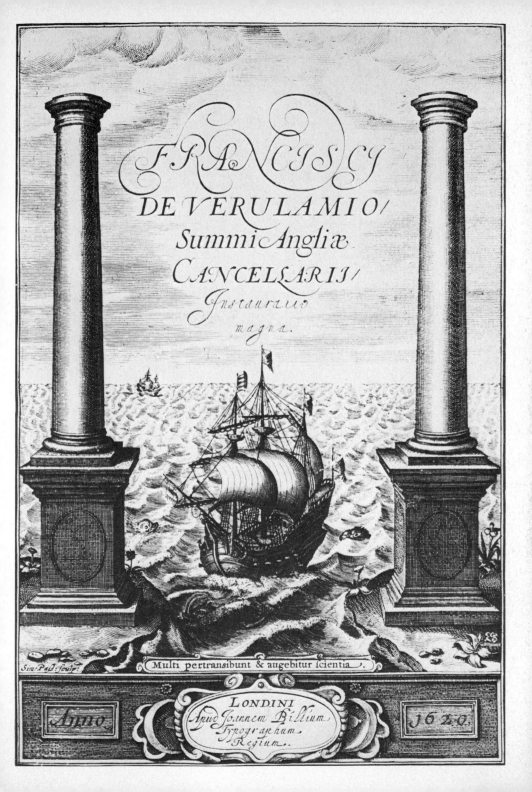

FRANCISCI
DE VERULAMIO,
Summi Angliæ
CANCELARIJ,
Instauratio
magna.

Sim: Pass: sculp:

Multi pertransibunt & augebitur scientia.

Anno

LONDINI
Apud Joannem Billium,
Typographum
Regium.

1620.

52   Francis Bacon (1561–1626), whose vision inspired the founders of the Royal Society and made him the prophet of the modern technological revolution.

be tested and verified by further experiments. Bacon believed that his scheme of systematic experiments would advance 'the true and lawful goal of the sciences . . . ; that human life be endowed with new discoveries and power.' That new power over nature would in its turn, he thought, make men more prosperous and thus happier. Bacon's belief in the achievement of scientific progress and material prosperity through systematic experiments upon natural phenomena made him the prophet of the technological revolution of the nineteenth and twentieth centuries.

Descartes, Bacon's contemporary, supplied the mathematical emphasis which was lacking in Bacon's theories but which was just as important as experiment and induction. The whole basis of Descartes's thinking was *deductive*. In his famous *Discours de la méthode* he started by doubting everything except the very fact of his own existence and proceeded to deduce the existence of God and of the whole material universe. He intended that each step in his argument should be as clear and certain as a mathematical proof.

These long chains of reasonings which geometers are accustomed to using to teach their most difficult demonstrations, had given me cause to imagine that everything which can be encompassed by man's knowledge is linked in the same way, and that provided only that one abstains from accepting any for true which is not true, and that one always keeps the right order for one thing to be deduced from that which precedes it, there can be nothing so distant that one does not reach it eventually, or so hidden that one cannot discover it.

Descartes's methodology, with its stress on deduction and on the importance of mathematics, was a perfect complement to Bacon's emphasis on experiment and induction.

Descartes and Bacon were the great early seventeenth-century publicists of new scientific methods, but even before they had produced their influential writings their ideas had been put into practice by Galileo. Galileo, combining the approaches of Descartes and of Bacon, applied to the study of natural phenomena a scientific method which has been recognized ever since as the correct way of studying nature and advancing scientific knowledge. One aspect of his methodology which was of the greatest significance was his ability, once he had observed and experimented upon the everyday phenomena of the world, to transcend these obvious natural realities and tackle scientific problems at a deeper level. For example, he rejected the idea that the motion of a body had any necessary connection with its weight; this enabled him to detach the motions of all bodies from the properties (such as size and weight) associated with these motions and to compare only the motions of the bodies. He was thus able to look for general laws of motion.

The modern scientific method created by the ideas of Bacon and Descartes and the work of Galileo can perhaps be summed up as follows: careful observation of and experiment upon the phenomena of the real world; the induction of general ideas from these observations and experiments; the testing of the general concepts so formed by deductions from them and by further experiments to verify these deductions; the application of precise measurements, involving the use of mathematics, during the experiments; and the ability to transcend the physical realities of the world and frame general concepts about the behaviour of bodies based on their fundamental properties, such as motion. Galileo, the man who put all these ideas into practice at one time or another in the course of his life, was the first 'modern' scientist, the first to apply recognizably modern scientific methods to the study of nature. This is perhaps the greatest of all his numerous claims to immortality, for the method which he was the first to apply has been the basic tool in all the spectacular advances in science which have taken place between the seventeenth century and our own day.

THE NEW INSTRUMENTATION

A third feature of the period which helped to make the scientific revolution possible was the creation of new scientific instruments. During the seventeenth century at least six crucially important scientific instruments were either invented or put to significant use for the first time; the telescope, the microscope, the thermometer, the barometer, the pendulum clock and the air pump. The air pump, which was invented by Otto von Guericke about the middle of the seventeenth century, enabled physicists to study the properties of the air in a logical and systematic way, and two notable Englishmen, Robert Hooke and Robert Boyle, both prominent members of the Royal Society, performed important experiments with air pumps. The pendulum clock was also invented at about the same time by the Dutchman Christiaan Huygens, who patented it in 1657. It made possible the measurement of small intervals of time which previously could be measured only inaccurately or not at all; this was of very great importance during a period when scientific advance was depending more and more on exact measurements of every kind. The barometer, invented in the

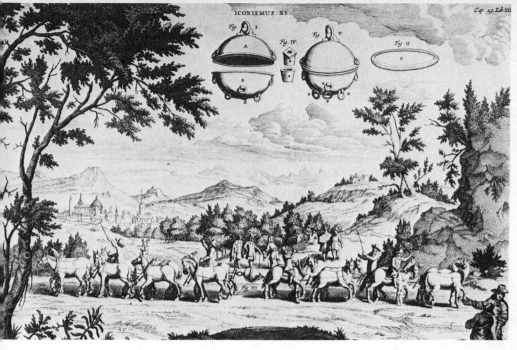

53, 54, 55 Von Guericke's experiment to demonstrate the power of a vacuum. Sixteen horses, divided into two teams, failed to pull apart a pair of hollow hemispheres that had been placed together and emptied of air. The air pump he invented is shown below left. One of Boyle's experiments, below right, involved putting a rat inside a container from which the air was then gradually exhausted.

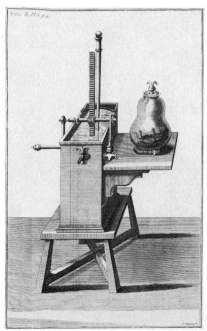

56 Two thermometers invented by the Italian doctor Santorio Santorio; the device on the right, driven by clockwork, was used for measuring the pulse rate.

early seventeenth century by Galileo's pupil Evangelista Torricelli, made possible the observation and measurement of variations in air pressure. The invention of the thermometer has been credited – perhaps wrongly – to Galileo, but the instrument was certainly in use in chemical studies in 1611.

The microscope, which greatly furthered the study of minute objects, had a long history, going back to the time of the ancient Greeks. Early microscopes, however, were all simple instruments with a single lens, and the compound microscope, consisting of multiple lenses, was not invented before the end of the sixteenth century, when it was probably developed in Holland. Galileo's detailed studies of insects were probably the first occasion on which a compound microscope was used for scientific purposes. The earliest large-scale treatise on microscopical observations was, however, that of Robert Hooke, who published his *Micrographia* in 1665. Four years earlier, the microscope had facilitated the confirmation of Harvey's theory of the circulation of the blood, the most important advance in physiological knowledge which had been made up to that time. Harvey had postulated the existence of capillaries which carried the blood from the finest arteries to the finest veins, thus completing the circulatory system. Capillaries, however, are in-

57 A thermometer made for the Accademia del Cimento, Florence, founded in 1657.

visible to the naked eye, and Harvey's inability to prove their existence delayed general acceptance of his theories. It was only in 1661 that the Italian Marcello Malpighi in his microscopical studies of the lungs of a frog identified capillaries.

The most significant of all the new instruments, in terms of its immediate impact on seventeenth-century science, was undoubtedly the telescope. It was probably invented in 1608 by a Dutch spectacle-maker of Middelburg, Hans Lippershey, but it was Galileo who used an improved version to open up a whole new dimension in astronomy. He has left his own account, written in 1610, of how he constructed his first telescopes.

About ten months ago a report reached my ears that a Dutchman had constructed a telescope by the aid of which visible objects, although at a great distance from the eye of

58, 59 Above, the head of an insect viewed through a microscope made during Galileo's lifetime, right.

the observer, were seen distinctly, as if near; and some proofs of its most wonderful performances were reported, which some gave credence to, but others contradicted. A few days after, I received confirmation of the report in a letter written from Paris by a noble Frenchman, Jaques Badovere, which finally determined me to give myself up first to inquire into the principle of the telescope, and then to consider the means by which I might compass the invention of a similar instrument, which after a little while I succeeded in doing, through deep study of the theory of refraction; and I prepared a tube, at first of lead, in the ends of which I fitted two glass lenses, both plane on one side, but on the other side one spherically convex, and the other concave. Then bringing my eye to the concave lens I saw objects satisfactorily large and near, for they appeared one-third of the distance off and nine times larger than when they are seen with the natural eye alone. I shortly afterwards constructed another telescope with more nicety, which magnified objects more than sixty times. At length, by sparing neither labour nor expense, I succeeded in constructing for myself an instrument so superior that objects seen through it appear magnified nearly a thousand times and more than thirty times nearer than if viewed by the natural powers of sight alone.

With his new telescope Galileo made in 1609 and 1610 a series of spectacular astronomical discoveries which he published in the latter year under the title *Sidereus Nuncius* (*The Starry Messenger*). The book rapidly became known throughout most of Europe and within five years of its publication it was being discussed as far away as Peking, evidence of the tremendous impact which these telescopic discoveries made on the human imagination. Galileo's and indeed all the earliest telescopes were of the refracting variety, in which the images for study were refracted or bent through a lens. The first reflecting telescope, the predecessor of all the largest telescopes of the

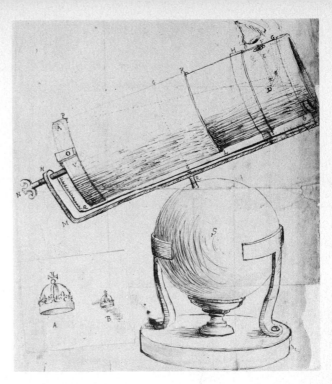

60 Newton's design for his
   reflecting telescope.

twentieth century, was constructed in 1668 by Isaac Newton,
who used a mirror to reflect the images of the heavenly bodies
which he was studying. Both refracting and reflecting telescopes,
and indeed all the contemporary scientific instruments, were
extremely primitive tools when compared with present-day
precision instruments; but at the time they added a series of
new dimensions to the study of nature and thus made a most
important contribution to the great scientific advances of the
period.

The scientists who began to use these new instruments from
the early seventeenth century onwards also helped to advance
scientific developments by a constant interchange of letters.
These sometimes anticipated and often supplemented their
books and thus helped to make their ideas available to fellow
scholars and, if the letters were published, to a wider public as
well. Such correspondence was not, of course, a new phenomenon, but during the seventeenth century it seems to have

been conducted on a much larger scale than ever before. In the earlier part of the period Marin Mersenne, a scientist who was also a Catholic priest, set himself up in Paris as a kind of scientific intelligencer, corresponding with the leading scientists of the day and passing new ideas from one to another. Such scientific correspondence continued to be of great importance during the latter part of the century and the letters of men as important as Newton and Leibniz contain ideas which cannot be found in their more formal works.

THE SCIENTIFIC SOCIETIES

During the second half of the seventeenth century, however, other methods of spreading scientific knowledge besides books and correspondence came to the fore with the creation of scientific societies and journals. The new societies were specifically designed to advance knowledge by promoting co-operative scientific work among their members. The most notable were the Royal Society of London and the Académie Royale des Sciences of Paris, both founded in the 1660s, but they had a short-lived predecessor in the Florentine Accademia del Cimento (Academy of Experiments), created in 1657. This was founded in the shadow of Galileo, who had died fifteen years before. The necessary financial support came from two of the Medici family who had studied under Galileo, the Grand Duke Ferdinand II and his brother Leopold, and the moving spirits on the scientific side were two of Galileo's most distinguished pupils, Evangelista Torricelli and Vincenzio Viviani. During the 1640s Ferdinand and Leopold had set up a well equipped scientific laboratory and in the early 1650s scientists began to meet there frequently for experiments and discussions. The creation of the Academy simply formalized this already existing society and during the ten years after 1657 its members did important work, especially in the field of physics. The researches were all carried out on careful experimental lines and a joint account was published by members of the society in 1666 under the title *Saggi di naturali esperienze fatte nell'Accademia del*

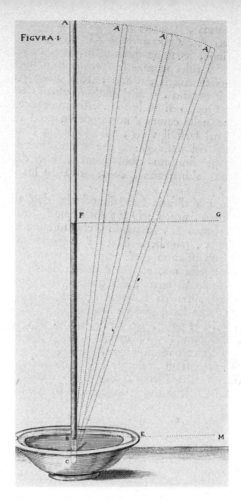

FIGVRA·I·

61   Barometer invented by Torricelli for use in his experiment demonstrating that air behaves according to mechanical laws.

Cimento. This contained particularly notable discussions of measurements of atmospheric pressure and of temperature which had been made with the aid of the barometer and the thermometer. In 1667, however, the Academy was discontinued. The reasons for this are not certain, but it may be significant that its dissolution took place during the same year that Prince Leopold was made a cardinal. He may have felt that continuing patronage of the society did not accord well with his new dignity; or perhaps this thought was suggested to him at the Papal Court!

83

62 The first charter of the Royal Society, granted by Charles II.

By that time the English Royal Society had developed from informal meetings of scientists which began at London and Oxford in the 1640s. It received its formal charter in the summer of 1662, some four and a half years before the foundation in France of the Académie Royale des Sciences. This, too, arose from informal scientific gatherings. A group of men who included Descartes, Blaise Pascal and Pierre de Fermat met in Paris during the 1650s to discuss scientific problems and suggest fresh researches, and finally, in December 1666, at the instigation of his adviser Jean-Baptiste Colbert, Louis XIV agreed to the establishment of a regular academy. The English and French institutions differed in significant ways. The Parisian Academy was very closely tied to the State, which appointed the Academicians and paid their salaries, whereas the Royal Society was completely independent of government control and its Fellows elected their own colleagues, who rapidly grew in numbers to well over a hundred, far more than the handful of French Academicians.

There were also, however, similarities between the two societies. In their early years both institutions laid some stress on the practical value of scientific researches. The Royal Society set up a committee for the history of trades, which concerned itself with industrial technology. The Académie des Sciences started a collection of tools and machines and examined new inventions which were submitted to it for approval. This Baconian concern for the practical uses of science did not last very long as far as the Royal Society was concerned, and if it did persist longer, in theory at least, in the work of the Académie des Sciences, this was largely because the latter was under firm government control. The Marquis de Louvois, the war minister, who took over supervision of the Academy after Colbert's death made it plain that he thought that scientific inquiries should be made to serve useful ends, by which he meant research 'that relates to the service of the king and the state'.

These efforts to harness science to practical ends produced little of significance in either Paris or London, and the theoretical work of both societies in such fields as biology, physics and astronomy was more important. Investigations in the last of these subjects were aided by the foundation of observatories at Paris and Greenwich in 1667 and 1675 respectively. The former was an offshoot of the Académie des Sciences, but the latter was only under rather ill-defined control by the Royal Society, a situation which led to a serious quarrel between the Society and the Astronomer Royal, John Flamsteed, in the early years of the eighteenth century. The specific contributions which the Royal Society and the Académie des Sciences made to scientific knowledge in the latter part of the seventeenth century were useful, but more important perhaps was the idea which they enshrined of science as a co-operative venture, an idea which only really took root in that century and which was to be of the very greatest importance in the years ahead.

This idea of co-operation among scientists was also aided by the establishment of scientific journals, the most important of which were the *Philosophical Transactions of the Royal Society*

63   A composite view showing work in progress at the Académie Royale des Sciences, Paris, in 1698. The range of activities is wide, with particular emphasis on the dominant sciences of astronomy, physics and mathematics.

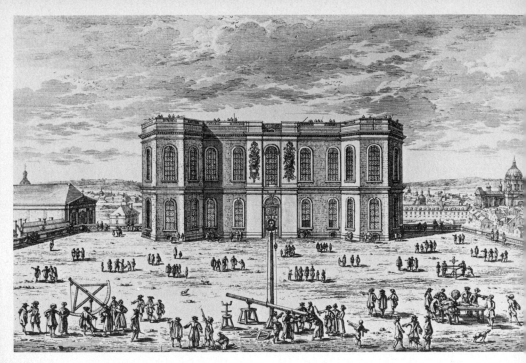

64　The Paris Observatory, a research centre for astronomy and physics, was built between 1667 and 1672.

and the *Journal des Sçavans*. The *Journal des Sçavans*, founded in January 1665 and published weekly, came to be an organ of communication not only among scientists themselves but also between scientists and an educated lay public, both in France and elsewhere in Europe, who were increasingly interested in scientific affairs. It published not only detailed results of experiments but also matters of general scientific interest and became the model for all subsequent scientific publications designed to appeal to a wide reading public. The *Philosophical Transactions*, first published in March 1665, only two months after the *Journal*, catered for a more specialized readership. The man behind it was Henry Oldenburg, the secretary of the Royal Society, and its substantial monthly volumes contained learned correspondence with foreign investigators and notices of new scientific books as well as papers contributed by members. It was,

65, 66 Greenwich Observatory: an anonymous picture painted a few years after
the building was completed in 1675. The contemporary ground plan, below,
indicates the modest scope of the establishment.

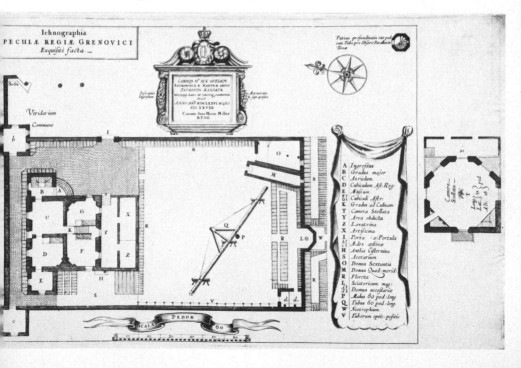

therefore, designed essentially for practising scientists rather than for a more general public and became a standard for the publications of later learned societies.

COPERNICUS

The new journals and societies and the correspondence of scientists themselves helped to spread scientific ideas widely throughout Europe during the course of the seventeenth century. The ideas which they propagated, taken together, constituted a revolution in man's understanding of the universe and of himself. The decisive role in the creation of these new ideas was played by a handful of men, the first of whom, Nicholas Copernicus, had no thought that he was initiating a scientific revolution when he published his life's work, *De Revolutionibus Orbium Coelestium* (*On the Revolutions of the Heavenly Spheres*), in 1543.

Copernicus was born in 1473, in the town of Toruń on the Vistula, the son of a prosperous merchant. At the age of eighteen he embarked on a long course of university studies, going first to Cracow in Poland, where he spent four years, and then on to Italy, to the universities of Bologna and Padua. His studies ranged widely, including philosophy, law and medicine as well as mathematics and astronomy, and in 1503 he took his degree as doctor of canon law at Ferrara. By that time he was already a canon of the cathedral of Frauenburg in East Prussia, a post to which he had been appointed by his uncle Lucas, who was Bishop of the diocese of Ermland, in which Frauenburg was situated. In 1506 he left Italy and returned home, spending the next six years with his uncle at Heilsberg Castle, residence of the Bishops of Ermland. In 1512, however, Bishop Lucas died and soon afterwards Copernicus, then a man of forty, took up his responsibilities as a canon of Frauenburg Cathedral, carrying them out for the remining thirty years of his life. The duties were not onerous, and he had plenty of time to devote to his astronomical work.

It seems to have been towards the end of his stay in Italy that the idea that the universe was heliocentric rather than geocentric (sun-centred rather than earth-centred) began to take root in his mind. This heliocentric idea was not, of course, new – it had been considered in the time of the Greeks and was much discussed in Italy itself while Copernicus was a student there – but it seems clear that, once Copernicus got hold of it, he clung on to it with unparalleled tenacity for the rest of his life. During the years between 1506 and about 1530, when he finished the manuscript of the *De Revolutionibus*, the details of his astronomical system must have been taking shape in his mind. Once he had completed his manuscript he laid it aside, making only occasional corrections to it. Copernicus's book was written in his grim and forbidding home in a three-storied tower of the fortified wall surrounding the cathedral of Frauenburg. From the second floor of the tower a door led out to a platform on top of the wall, from which he had an open view of the Baltic Sea to the north and an extensive plain to the south, as well as a spectacular panorama of the stars at night. He was not, however, a great observational astronomer – he only wrote down between sixty and seventy observations during his entire lifetime – and his work was founded on the records left by his predecessors.

The *De Revolutionibus* remained in manuscript for many years after its completion. Copernicus was a diffident man who tended to undervalue his own work and he feared that publication would expose him to the ridicule of fellow astronomers. In 1542, however, he was at last persuaded to send the book to the printer, and the first completed copy arrived at Frauenburg in May 1543, just before his death.

Copernicus himself would have been astounded if he could have known about the long-term significance of his work. The 'timid canon', as a recent biographer has aptly christened him, was essentially a conservative figure. Throughout his life he accepted unquestioningly the basic principles of Aristotelian physics and the axioms of uniform circular motion in the

heavens and of the existence of the heavenly spheres. His attack on the Ptolemaic system in astronomy was due to dissatisfaction with some of its technical aspects. He pointed out, for example, that, to be worthwhile, the calendar reform which was under serious discussion in the early sixteenth century would have to be preceded by astronomical reform, and he hoped that his own work might help to make a new calendar possible. More important was the fact that Ptolemy's complicated geocentric system of epicycles, eccentrics and equants, and its later developments, did not fit in exactly with the observed motions of the heavenly bodies, and Copernicus hoped that his own heliocentric system would reconcile theory and practice by accounting precisely for observed celestial motions. Above all, Copernicus objected to Ptolemy's use of the equant, which meant that in the latter's system the heavenly bodies moved with circular motion around one centre and with uniform speed around another. This, Copernicus thought, was a fundamentally unreasonable way of accounting for the generally accepted axiom of uniform circular motion, which, he argued, could mean only one thing, uniform circular motion around a single point.

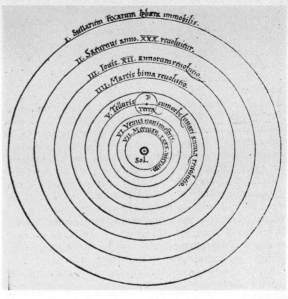

The *De Revolutionibus* is divided into six books, and the greater
part of the work is taken up by elaborate mathematical and
astronomical calculations. The basic principles are, however,
set out in the first book, and may be summarized as follows. The
universe, Copernicus believed, consisted of eight spheres – the
motions which he assigned to the earth enabled him to dispense
with the ninth and tenth spheres – with the sun at rest at the
centre and the sphere of the fixed stars, at the boundary of the
universe, also at rest. The planets, Mercury, Venus, the earth,
Mars, Jupiter and Saturn, revolved round the sun in that order,
while the moon revolved around the earth. The apparent
movements of the stars and the sun were due to the daily
rotation of the earth on its axis and to its annual journey round
the sun. When, however, Copernicus embarked on the detailed
calculations which were necessary to prove his ideas, he ran
into difficulties, and in the body of the work he was forced
to modify the clear and simple theory, which he had set out
in the first book, to such an extent that his final picture was
hardly, if at all, less complicated than Ptolemy's. He did manage
to get rid of equants in his calculations, but retained epicycles
and eccentrics and found that he was unable to keep the sun

93

at the exact centre of any of the planetary circles. The sun, it is true, remained at the centre of the motionless sphere of the fixed stars, but the planets, including the earth, all revolved around a point in space which was some distance away from the sun.

Copernicus, therefore, did not really succeed in his objective, that of replacing the illogical and overcomplicated Ptolemaic system with a logical and simple structure of his own. His work did, however, have a profound long-term significance which was appreciated during the seventeenth century, when his heliocentric ideas became generally accepted in the astronomical world. Copernicus himself never questioned the existence of the traditional crystalline spheres, but his work, in fact, destroyed the main reason for believing in them. As long as it was thought that the stars revolved around the earth it was difficult to find an alternative to the hypothesis that they were all fixed to a revolving sphere: that theory alone could rationally account for the uniform pattern they always displayed in their motion around the earth. Copernicus's theory, however, reduced the stars to rest. This meant that there were only the individual motions of the earth and planets to explain, and this could be done perfectly well without introducing the idea of crystalline spheres. Thus, Copernicus's system put in jeopardy the whole two-thousand-year-old mechanism of the universe, founded on the existence of spheres, although it was left to Tycho Brahe and Kepler to demonstrate in a positive way, at the turn of the sixteenth and seventeenth centuries, that they did not exist.

Copernicus's work also raised the possibility that the universe was infinite. If his heliocentric theory were correct, the position of the stars as observed from the earth might have been expected to change slightly during the course of the year as the earth travelled round the sun. No such annual parallax could, however, be detected, and Copernicus solved the problem by arguing that the stars were enormously distant, far further from the earth than anyone had previously believed possible. He himself did not push the argument to its logical conclusion, but others

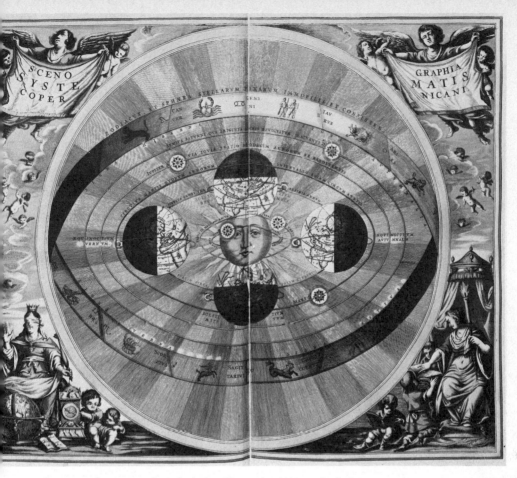

70   Seventeenth-century chart depicting Copernicus's cosmological system.

did. One was the daring philosopher Giordano Bruno who realized very clearly that the Copernican theory, by making the existence of the spheres unnecessary and removing the stars to unimaginable distances, made possible the conception of an infinite universe.

Copernican ideas had profound consequences for physics no less than for astronomy. Just as Copernicus himself accepted the concept of the crystalline spheres, so he also accepted the axioms of Aristotelian physics; but as well as making the spheres

71 An armillary sphere, made as late as 1552, showing the traditional, pre-Copernican structure of the universe, with the earth at the centre.

unnecessary, his astronomical system also rendered the maintenance of traditional ideas about physics impossible. The whole of Aristotelian physics depended on a basic difference between the motions of bodies on a stationary earth and the motions of the heavenly bodies. But Copernicus's theory set the earth moving among the other planets and therefore, at one stroke, abolished Aristotle's basic distinction. It was not until the early years of the century, however, that Galileo was able to put forward ideas about motion which made the Copernican system seem a genuine physical as well as an astronomical possibility, and it is significant that it was only after Galileo's work that Copernican ideas began to gain general acceptance among scientists.

Copernicus's theory had important religious implications as well as profound scientific consequences. If, for example, the earth as a planet and therefore a celestial body partook of the pure nature of the heavens, how could the heavens, which in turn must partake of the nature of the planetary earth, with all its evils and imperfections, be a suitable home for God? Moreover, if the universe were infinite, as Copernicus's theory implied, where was God's abode situated? It was some time after the publication of the *De Revolutionibus* before such questions began to be asked and even longer before they began to

everything moving

be answered, but there was an immediate and hostile reaction to Copernicanism from the leaders of the Protestant churches. Despite the fact that his book was not published until 1543, his ideas were known in intellectual circles in Europe well before that, and in the 1530s Martin Luther made a contemptuous reference to Copernicus as 'the new astrologer who wants to prove that the earth moves and goes round. . . . The fool wants to turn the whole art of astronomy upside down.' However, Luther added, 'as Holy Scripture tells us, so did Joshua bid the sun stand still and not the earth.' Melanchthon, Luther's principal lieutenant, also berated Copernicus in 1549:

> The eyes are witnesses that the heavens revolve in the space of twenty-four hours. But certain men, either from the love of novelty, or to make a display of ingenuity, have concluded that the earth moves, and they maintain that neither the eighth sphere [of the fixed stars] nor the sun revolves. . . . Now it is a want of honesty and decency to assert such notions publicly, and the example is pernicious. It is the part of a good mind to accept the truth as revealed by God and to acquiesce in it.

Calvin also joined in the condemnation in his *Commentary on Genesis*, where he cited the opening verse of the ninety-third psalm, 'the world also is stablished that it cannot be moved', and asked, 'who will venture to place the authority of Copernicus above that of the Holy Spirit?'

These vehement condemnations by Protestant leaders were not repeated at the time by Catholic dignitaries, who were less immediately ready to resort to literal interpretations of Scripture than their Protestant counterparts, and it was not until 1616, following the speculations of Bruno and the work of Galileo, that Copernicanism was declared false by the Holy Office. This condemnation came just when it was beginning to be generally acceptable to astronomers, and by the middle of the seventeenth century, despite ecclesiastical opposition, almost all important astronomers were Copernicans.

97

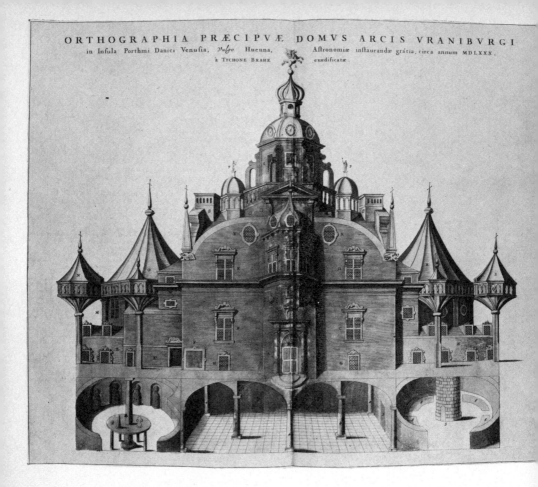

ORTHOGRAPHIA PRÆCIPVÆ DOMVS ARCIS VRANIBVRGI
in Insula Porthmi Danici Venusia, *vulgo* Huenna,   Astronomiæ instaurandæ grätia, circa annum MDLXXX,
à Tychone Brahe   exædificatæ.

TYCHO BRAHE

Tycho Brahe, the next great astronomer of the period, was the scion of a distinguished Danish noble family. Born in 1546, it was hoped that he would become a statesman, thus following family tradition. When he was only thirteen he was sent to the university of Copenhagen to study rhetoric and philosophy in preparation for his proposed career in public life. Within a year, however, he witnessed a partial eclipse of the sun which made an overwhelming impression on him and determined the whole future course of his life: he decided to be an astronomer. It was the predictability of the eclipse that really impressed the young

72   Opposite: Tycho Brahe's main observatory at Uraniborg, built between 1576 and 1580.

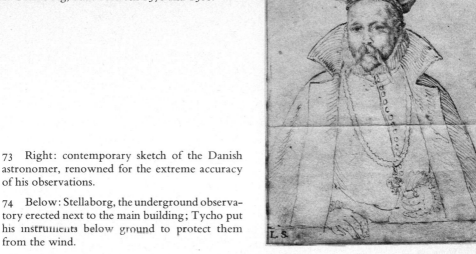

73   Right: contemporary sketch of the Danish astronomer, renowned for the extreme accuracy of his observations.

74   Below: Stellaborg, the underground observatory erected next to the main building; Tycho put his instruments below ground to protect them from the wind.

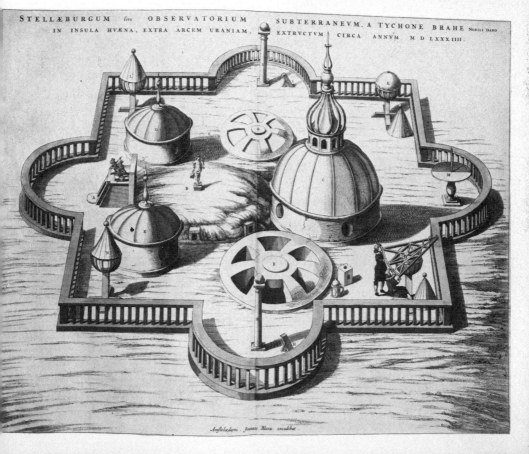

STELLÆBURGUM sive OBSERVATORIUM SUBTERRANEVM, A TYCHONE BRAHE Nobili Dano IN INSULA HVÆNA, EXTRA ARCEM URANIAM, EXTRVCTVM CIRCA ANNVM M D LXXXIIII.

Amstelædami . Joannis Blaeu excudebat .

Tycho. It struck him as 'something divine that men could know the motions of the stars so accurately that they were able a long time beforehand to predict their places and relative positions'. He immediately began to buy astronomical books and during the remaining years of his university studies, which continued at Leipzig, Wittenberg, Rostock, Basle and Augsburg, he began to design bigger and better instruments for observing the skies.

In 1572, soon after Tycho's return to Denmark at the end of his studies, a 'new star' appeared in the sky, where none had previously been visible, and his observations of this remarkable phenomenon, which he published the following year, immediately made him the most famous astronomer in Europe. Five years later his observations of the comet of 1577 dealt further blows to Aristotelian cosmology. By that time he had already begun to establish his remarkable observatory at Uraniborg, on the island of Hveen, which had been granted to him in May 1576 by King Frederick II of Denmark. In 1576 Tycho had been planning to settle permanently in Basle, but the King was determined to keep Europe's leading astronomer at home and offered him Hveen, situated in the Sound, between Copenhagen and Elsinore, 'to have, use and hold, quit and free, without any rent, all the days of his life, and as long as he lives and likes to continue and follow his *studia mathematices*.' This magnificent gift was accompanied by an annual grant and various sinecures, which gave Tycho one of the highest incomes in Denmark. He worked on the island until 1597, when he left, following quarrels with Frederick's successor, Christian IV, to journey restlessly through central Europe until 1599; then he settled down near Prague as imperial mathematician to the Emperor Rudolph II. He never effectively resumed his observations, however, and the main interest of the last years of his life was his meetings with Kepler, which began in 1600 and lasted for eighteen months, until Tycho's death in October 1601.

Tycho was a most remarkable man both in character and appearance. While still a student he lost the bridge of his nose in a duel with another Danish youth. He had the missing piece replaced with a construction of gold and silver alloy, and this strange feature, especially when combined with his handlebar moustache, egg-shaped head and haughty manner, must have made him a formidable sight. He was, indeed, a very arrogant man and treated his tenants in Hveen in a most high-handed way, exacting from them services to which he was not entitled and imprisoning them when they refused to comply. Such intolerable conduct was one of the reasons for his disputes with Christian IV, and it is a contradiction in Tycho's character that he combined unreasonable arrogance in his dealings with his fellow men with great humility towards the facts of science.

Tycho's great contribution to astronomy was, as his biographer, J. L. E. Dreyer, puts it, the realization 'that only through a steadily pursued course of observations would it be possible to obtain a better insight into the motions of the planets'. During his twenty-year residence on Hveen he made a series of precise observations of the movements and positions of the planets and stars which had no previous parallel in the history of astronomy. Unlike earlier astronomers who had been content to take observations of the planets at specific points in their orbits and then interpolate the rest of the orbits between these points, Tycho and his assistants followed the exact paths of the planets nightly for twenty years and carefully recorded the results. This thoroughness, together with the great size and accuracy of Tycho's instruments, made his data very much more reliable than any previous observations, and it was only the exceptional precision of his methods which enabled him to draw with confidence his extremely important conclusions about the nova of 1572 and the comet of 1577.

Tycho set out his conclusions about the new star in his *De Nova Stella*, written in the following year.

Last year in the month of November, on the eleventh day of that month, in the evening after sunset, when, according to my habit, I was contemplating the stars in a clear sky, I noticed that a new and unusual star, surpassing the other

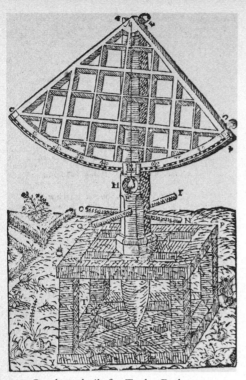

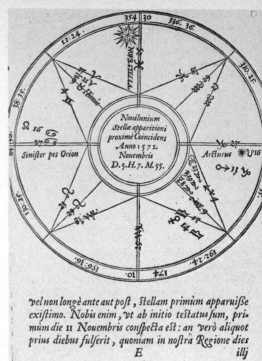

vel non longè ante aut poſt, ſtellam primùm apparuiſſe
exiſtimo. Nobis enim, vt ab initio teſtatus ſum, primùm die II Nouembris conſpecta eſt: an verò aliquot
prius diebus fulſerit, quoniam in noſtra Regione dies
E   illj

75 Quadrant built for Tycho Brahe at
Augsburg in 1569.

76 Woodcut from *De Nova Stella*, showing the positio
of the supernova of 1572.

stars in brilliancy, was shining almost directly above my head;
and since I had, almost from boyhood, known all the stars of
the heavens perfectly (there is no great difficulty in attaining
that knowledge), it was quite evident to me that there had
never before been any star in that place in the sky, even the
smallest, to say nothing of a star so conspicuously bright as
this.

By careful observations and calculations he reached the conclusion that 'this new star is neither in the region . . . below the
moon nor among the orbits of the seven wandering stars, but
it is in the eighth sphere, among the other fixed stars.' The 'new
star' was in fact a very distant exploding star, but at the time
Tycho's conclusion was sensational, a direct challenge to the

7   Engraving of the great mural quadrant at Uraniborg; the observer is Tycho.

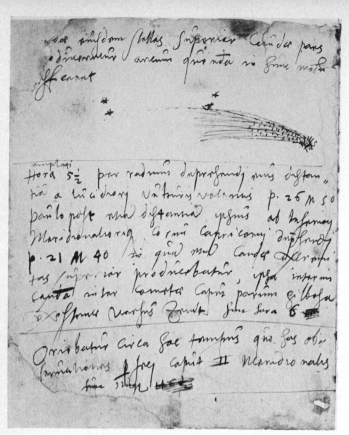

78 Tycho Brahe's notes on
the comet of 1577.

fundamental Aristotelian idea of an unchanging, perfect
universe above the sphere of the moon.

Tycho's second set of observations, of the comet of 1577,
convinced him that it was not a sublunary object, as comets
were generally supposed to be, but that it was moving in an
unhindered manner through the regions between the planets.
According to Aristotelian doctrine this was impossible, partly
because the regions beyond the moon were supposed to be
unchanging, and partly because such a comet would have to
pass through the crystalline planetary spheres. Copernicus's
ideas had made the spheres unnecessary: now Tycho's observations showed that they could not exist.

Tycho was not content to spend his life as a mere observational astronomer, however important he held that work to

be. He constructed his own system of the universe as an alternative to the Ptolemaic and Copernican theories. According to Tycho all the planets revolved around the sun and the entire group of sun and planets then revolved around the earth-moon system. This theory enjoyed a wide popularity in the early seventeenth century among those who thought the Ptolemaic system out of date but found it difficult to accept the central Copernican idea of a moving earth. Tycho's lasting claim to fame does not, of course, rest on this interesting but mistaken compromise system, but on the fact that he provided the data without which Kepler could never have fashioned his planetary laws.

79 Diagram of the universe according to Tycho Brahe, who did not accept the Copernican idea of a sun-centred solar system.

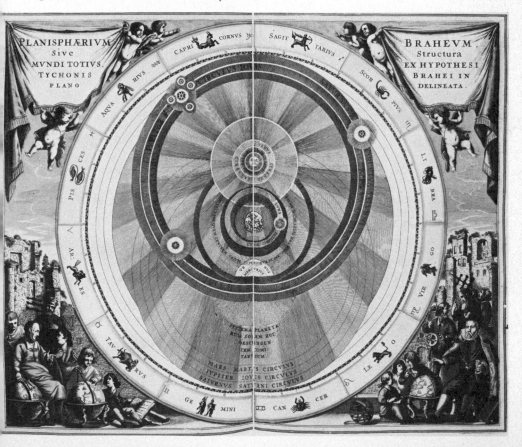

80 Johannes Kepler, who combined the speculations of Copernicus with the observations of Brahe to discover the laws governing the orbits of the planets.

Johannes Kepler was born in December 1571 in the town of Weil in south-west Germany. His grandfather, Sebaldus Kepler, was said to be of noble descent and became Mayor of Weil, but Sebaldus's offspring were, as described by Arthur Koestler, 'mostly degenerates and psychopaths who chose mates of the same ilk'. Johannes's father was an adventurer who almost ended his career on the gallows and his mother was nearly burned as a witch. Young Kepler seems to have had a ghastly childhood among his horde of quarrelsome and unattractive relations, and his spirits cannot have been helped by the fact that he suffered constantly from ill-health.

Kepler's parents destined him for the Lutheran ministry and after irregular studies at elementary school he entered a theological seminary when he was thirteen years old in preparation for study at the university of Tübingen, where he became a student four years later. He graduated in arts there when he was twenty and then matriculated in the theological faculty, where he studied for nearly four years. He seems, however, to have become unable to accept all the rigidities of Lutheran doctrine. Under the influence of Tübingen's distinguished astronomer, Michael Maestlin, he developed a great interest in astronomy and considerable aptitude for mathematics. In these circumstances he decided to abandon divinity before he sat his final examinations and to accept a post as teacher of mathematics and astronomy in Graz.

Kepler was in many ways a medieval figure. He believed in astrology and in an extreme form of neo-Platonic mathematical mysticism which led him to elaborate the idea that the universe was built around certain geometrical figures, such as the pyramid and the cube. His duties at Graz were sufficiently light to leave him time to write up these ideas, and his book on the subject, *Mysterium Cosmographicum*, was published in 1596, when he was twenty-five. He sent copies to distinguished scientists, among them Tycho Brahe, who also believed in astrology and was not put off by the mysticism of a book which incidentally revealed that Kepler was exceptionally able at astronomical computations. He therefore invited Kepler to join him in his work; Kepler finally took up the offer in 1600.

The partnership thus established was far from smooth during the eighteen months it lasted, but, on Tycho's death, Kepler got his reward. He took possession of Tycho's astronomical papers and held on to them despite the complaints of the latter's heirs. It was these data that enabled him to discover the three laws of planetary motions. He was appointed Tycho's successor as imperial mathematician by Emperor Rudolph, but in 1612 Rudolph died and Kepler took up the post of mathematician at Linz, where he remained for fourteen years. During the last few years of his life he had no fixed home. He died at Regensburg in 1630.

Kepler's great contribution to science lay in his laws of planetary motions, which he discovered after tremendous and protracted intellectual struggles and elaborate mathematical calculations. He published the first two in his *Astronomia Nova* of 1609 and the third in *Harmonice Mundi* in 1619. Kepler had accepted Copernicus's heliocentric views while still a student, but in his first law he demonstrated that the planets' orbits were not circular, as Copernicus and all his predecessors right back to Aristotle had assumed, but elliptical, with the sun at one focus of the ellipse. In his second law, which he actually discovered before the first, he overthrew another of the fundamental Aristotelian tenets that Copernicus had shared – the idea that the motion of the planets was uniform. He demonstrated instead that the speed of a planet in its orbit falls off in proportion as its distance from the sun increases, in such a way that if a line could be drawn joining the planet to the sun that line would sweep out equal areas of space in equal times.

Kepler's third law, published ten years later, showed that the squares of the times taken by any two planets in their revolutions around the sun were in the same ratio as the cubes of their average distances from the sun. An example may make this clear. If we take the earth and Saturn as our two planets,

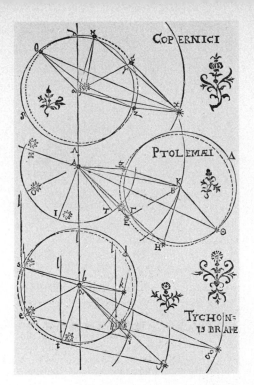

81 Three conflicting theories – the orbit of Mars according to Copernicus, Ptolemy and Brahe: diagram from Kepler's *Astronomia Nova*.

then the earth's distance from the sun is our unit of distance and the earth's year becomes our unit of time. Saturn's average distance from the sun is well over nine units, but we will assume that the figure is exactly nine in order to simplify our calculation. If we represent the time Saturn takes to revolve around the sun by the letter x then, according to Kepler's third law, 1^2 and x^2, the squares of the times taken by the earth and Saturn respectively in their orbits are in the same ratio as 1^3 and 9^3, the distances of the earth and Saturn from the sun. 1^3 is 1, 9^3 is 729, and x is therefore equal to the square root of 729, which is 27. Saturn's year is therefore equal to 27 earth years. This is correct on the basis of our simplified calculation, though Saturn's actual distance from the sun, nearly 9·7 times that of the earth, means that in fact it takes about 30 years in its orbit.

Kepler's discoveries were of immense significance. Before his work there was no wholly compelling reason for grouping the planets into a separate community independent from the

Harmoniæ Planetarum quinque, diſſimulatâ Venere						
Generis Duri				**Generis Mollis**		
In Tenſione Graviſſima.		Acutiſſima.		In Tenſione Graviſſima.		Acutiſſima.
	Sc. pr. ſec.	Sc.Pr.Se.			Sc.pr.ſec	Sc.Pr.Se.
d ♄♃	342. 18	351. 24.		d ♄♃	342. 18.	351. 24.
♄♃	285. 15	292. 48.		♄♃	273. 50	280. 57.
g ♃	228. 12	234. 16.		g ♃	228. 12.	234. 16.
d ♃	171. 9	175. 42.		d ♃	171. 9.	175. 42.
Venus hic ob- c♀ ſtrepit	95. 5	97. 37.		Venus ob ſtrepit c♀	95. 5	97. 37.
Ter. g iiij	7. 3	58. 34.		Ter. g iiij	57. 3.	58. 34.
♂ biiij	35. 39.	36. 36.		♂ biiij	34. 14.	35. 8.
giij	28. 31.	29. 17.		giij	28. 31.	29. 17.
♃ dj	5. 31.	5. 30.		♃ dj	5. 11.	5. 30.
bj		4. 35.				
♄ ♭ G.	2. 13.			♄ ♭ G	2. 8.	2. 11.
	1. 47.				1. 47.	1. 50.

Hîc in graviſſima tenſione concurrunt Saturnus, Terra, aphelijs; in media Saturnus perihelio, Jupiter aphelio; in acutiſſimâ, Jupiter perihelio:

Hîc non toleratur aphelius Jovis, at in acutiſſima tenſione concurrit Saturnus perihelio proximè.

82 Table from Kepler's *Harmonice Mundi*, relating the movements of the planets.

fixed stars, though Copernicus's theories, with their implications about the immense distances of the stars, clearly made such an idea tenable. Kepler's third law, however, demonstrated that the solar system was a unit on its own, containing planets whose orbits were mathematically related to one another. By thus revealing the internal mechanism of the planetary system he effectively separated it from the inaccessible stars. Moreover, through his first and second laws he showed that the idea of uniform circular motion, which had dominated astronomical thought for two thousand years, was a myth – the speed of each planet varied at different points in its elliptical orbit around the sun. This conclusion also confirmed Tycho's work in demonstrating that the supposed crystalline spheres, which formed the framework of the traditional universe, did not exist. Finally, Kepler's laws formed an essential element in Newton's great synthesis of 1687.

GALILEO

Kepler was nearly eight years younger than his even greater contemporary Galileo Galilei, who was born in Pisa early in 1564, the son of a poor but noble family. When he was eleven years old he was sent to a Jesuit school and by the age of seventeen had become a novice in the order. His father, however, was opposed to his entering the religious life and it was eventually decided that he should go to the university of Pisa. He wanted to study mathematics, but mathematicians were poorly paid and his father insisted that he should devote himself to medicine, which might be expected to bring him substantial monetary rewards after graduation. From 1583 onwards, however, Galileo devoted more and more of his time to mathematics and his father at last reluctantly gave him permission to abandon his medical studies. During the remainder of the 1580s Galileo worked hard at problems of mechanics, drawing conclusions based on his own experiments. He did not publish anything, but as he circulated his results in manuscript he acquired something of a reputation and in 1589, at the age of twenty-five, was appointed professor of mathematics at Pisa. Three years later he moved to a chair at Padua, one of the most famous of all the European universities. There he continued his researches into mechanical problems and, although he still published nothing, the important work he did in the 1590s bore fruit much later.

Galileo first became universally famous in 1610, at the time of the publication of his *Sidereus Nuncius*, which revealed him as a protagonist of the Copernican system. He defended the system explicitly three years later in a study of sunspots which contained a reference to 'the great Copernican system, to the universal revelation of which doctrine propitious breezes are now seen to be directed towards us, leaving little fear of clouds or crosswinds.' Galileo had been a convinced Copernican since his twenties but had been afraid to state his convictions for fear of ridicule. Now at last, in his late forties, he came out into the open. In 1610, after publishing his astronomical discoveries, he went to live in Florence as mathematician to the Grand Duke

of Tuscany, Cosimo de' Medici, but he soon came under attack from the Dominican order, which had a vested interest in the maintenance of Aristotelian ideas – St Thomas Aquinas himself had been a Dominican. Galileo had some influential ecclesiastics on his side, but Cardinal Bellarmine, the Church's leading theologian, saw only the danger of a scandal which might weaken Catholicism in its struggle with Protestantism, and it was decided to close the issue by declaring Copernicanism erroneous. In February 1616 Galileo was told that he must neither hold nor defend the doctrine, though he could still discuss it as a mathematical supposition, and in March of the same year it was officially condemned in a Holy Office decree.

In 1624 Galileo went to Rome in the hope of obtaining a reversal of the decision of 1616. He had six long audiences with Pope Urban VIII and although he did not succeed in getting any change in the Church's official attitude, the Pope did grant him permission to write about the 'systems of the world', both Ptolemaic and Copernican, as long as he came to a conclusion dictated in advance by the Pope: namely, that we cannot presume to know the structure of the universe because this would restrict God's omnipotence. The result was Galileo's *Dialogo . . . sopra i due Massimi Sistemi del Mondo* (*Dialogue on the Two Chief Systems of the World*), brilliantly written and published in Italian in 1632. It was widely acclaimed as both a literary and a scientific masterpiece, but soon led to Galileo's trial by the ecclesiastical authorities.

The *Dialogue* was in the form of a discussion between Salviati, a supporter of Copernican ideas; Simplicio, a good-natured but rather stupid supporter of Aristotle and Ptolemy; and Sagredo, an open-minded amateur. In the course of the work Galileo ruthlessly revealed over and over again the ignorance and confusion of Simplicio, and in these circumstances the Pope's dictated conclusion, especially when it was made to come from Simplicio's mouth, carried little conviction! The *Dialogue* was rightly seen as a Copernican tract, and the Pope, understandably angry, ordered the trial of Galileo, who was forced to travel to Rome in 1633 accused of 'vehement suspicion of heresy'. He was found guilty and forced to 'abjure, curse and detest' his errors. He was also condemned to imprisonment during the Holy Office's pleasure, but this sentence was immediately commuted by the Pope to house arrest at Galileo's own estate at Arcetri near Florence, where he remained until his death in 1642. These last years of his life were among the most productive of his entire career. He took up again his studies of motion and published the results in 1638 in admirably lucid prose in his *Discorsi e dimostrazioni intorno a due nuove scienze* (*Discourses . . . on Two New Sciences*).

Galileo's work touched seventeenth-century physical science at most of its important points. The most significant of all his contributions was almost certainly the first creation of a recognizably modern scientific methodology, but the details of his researches in mechanics and astronomy were also of the very greatest importance. Galileo's essential contribution to mechanics probably lay in his studies of acceleration and his discovery of the idea of inertia, where he both built on and modified the theories of his medieval predecessors. He described in his *Two New Sciences* an experiment in which he showed that the uniform acceleration which Heytesbury and his successors had discussed and defined in the fourteenth century could be identified experimentally with motion of a familiar kind. This experimental approach typifies perfectly the difference between Galileo and medieval philosophers in their approach to mechanics. Whereas medieval scholars were preoccupied with theory, Galileo got down to experiments on the realities of nature. In doing so he gave an original and vastly important new slant to the whole science of mechanics. Here, in his own words, is the famous 'acceleration experiment'.

A piece of wooden moulding . . . about twelve cubits long, half a cubit wide, and three fingerbreadths thick, was taken; on its edge was cut a channel a little more than one finger in breadth. Having made this groove very straight, smooth and polished, and having lined it with parchment, also as smooth and polished as possible, we rolled along it a hard, smooth and very round bronze ball. Having placed this board in a sloping position by lifting one end some one or two cubits above the other, we rolled the ball, as I was just saying, along the channel, noting, in a manner presently to be described, the time required to make the descent. We repeated this experiment more than once in order to measure the time with an accuracy such that the deviation between two observations never exceeded one-tenth of a pulse beat. Having performed this operation and having assured ourselves of its reliability, we now rolled the ball only one-quarter the length of the channel; and having measured the time of its descent, we found it precisely one-half of the former. Next we tried other distances, comparing the time for the whole length with that for the half, or with that for two-thirds, or three-fourths, or indeed for any fraction; in such experiments, repeated a full hundred times, we always found that the spaces traversed were to each other as the squares of the times,

84 Frontispiece of Galileo's *Dialogue on the Two Chief Systems of the World*, showing Aristotle, Ptolemy and Copernicus engaged in scientific debate; in the actual book discussion is carried on by three characters created by Galileo himself. ▶

DIALOGO
di
GALILEO GALILEI LINCEO
AL SER.mo FERD. II. GRAN. DVCA DI
TOSC   ANA

Stefan. Della Bella. F.

and this was true for all inclinations of the plane, i.e. of the channel along which we rolled the ball. . . .

. . . For the measurement of time, we employed a large vessel of water placed in an elevated position; to the bottom of this vessel was soldered a pipe of small diameter giving a thin jet of water, which we collected in a small glass during the time of each descent, whether for the whole length of the channel or for a part of its length; the water thus collected was weighed, after each descent, on a very accurate balance; the differences and ratios of these weights gave us the differences and ratios of the times, and this with such accuracy that although the operation was repeated many, many times, there was no appreciable discrepancy in the results.

Galileo had already reasoned that the character of motion on an inclined plane was the same as that of a body in free fall, so what he had done, though he himself did not put it in these terms, was to show that a uniform force (in this case the force of gravity) acting on a body produced a uniform acceleration. This was, in effect, an experimental demonstration of the central idea in Newton's second law of motion.

Galileo also discovered the idea of inertia. He came to agree with Buridan's contention that impetus was permanent – that a moving body would continue in motion for ever unless prevented from doing so by some external force. In his *Two New Sciences*, however, he saw impetus as a measure rather than a cause of motion. In other words, impetus lost its position as an intermediary cause of motion between the original motive force and the object moved, leaving the familiar Newtonian doctrine that a continually acting force itself causes acceleration. But what if the force causing acceleration were suddenly removed? Galileo argued in the *Two New Sciences* that 'any velocity once imparted to a moving body will be rigidly maintained as long as the external causes of acceleration or retardation are removed.' This summarizes the idea of inertia, the idea that a state of uniform motion is just as 'natural' as a state of rest.

Before Galileo formulated this concept, scientists had sought to explain motion itself. Afterwards, their task was to explain changes in motion. This was a fundamental alteration in outlook and one of the most important developments in the scientific thought of the time. Galileo's ideas about inertia did not correspond exactly with the classical law on the subject as it was stated later by Descartes and later still by Newton in his first law of motion, the idea that inertial movement continued in a straight line. Galileo believed that inertial movement, at least on a large scale, was circular, a belief that fitted in well with his general views about circularity in the heavens – he ignored Kepler's researches and continued to believe that the planets revolved in circular orbits around the sun. This error of Galileo's is, however, of slight importance compared with the significance of his discovery of the concept of inertia and the general importance of his work in mechanics. Historians of science agree that that work was of very great significance, but there are important differences of interpretation between those who see his work on motion as essentially the culmination of a medieval tradition, and those who stress its new features, which pointed the way to Newton and the classical laws of dynamics. It is surely best to conclude that each of these interpretations tells an important truth: Galileo certainly rounded off and completed a medieval tradition in mechanics, but he was also the herald of new ideas which were to hold the field until the Einsteinian revolution of the twentieth century.

Although Galileo's work on mechanics was of the most profound importance it was much less spectacular than his astronomical observations, and it was these which had by far the greater immediate impact. After describing, in his *Sidereus Nuncius*, how he constructed his first telescopes, Galileo went on to discuss the wonders which these instruments revealed.

Let me first speak of the surface of the moon which is turned towards us. For the sake of being understood more easily, I distinguish two parts in it, which I call respectively the

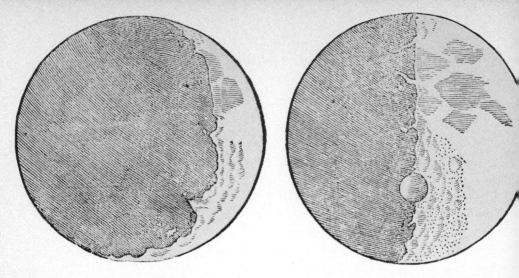

85, 86 Sketches from Galileo's *Sidereus Nuncius*. Left, the three-day-old moon; right, the moon at its first quarter.

brighter and the darker. The brighter part seems to surround and pervade the whole hemisphere, but the darker part, like a sort of cloud, discolours the moon's surface and makes it appear covered with spots. Now these spots, as they are somewhat dark and of considerable size, are plain to everyone, and every age has seen them, wherefore I shall call them great or ancient spots to distinguish them from other spots, smaller in size, but so thickly scattered that they sprinkle the whole surface of the moon, but especially the brighter portion of it. These spots have never been observed by anyone before me; and from my observations of them, often repeated, I have been led to that opinion which I have expressed, namely, that I feel sure that the surface of the moon is not perfectly smooth, free from inequalities, and exactly spherical, as a large school of philosophers considers with regard to the moon and the other heavenly bodies, but that, on the contrary, it is full of inequalities, uneven, full of hollows and protuberances, just like the surface of the earth itself, which is varied everywhere by lofty mountains and deep valleys. . . . The grandeur, however, of such prominences and depressions in the moon seems to surpass both in magnitude and extent the ruggedness of the earth's surface.

Galileo then turned his attention to the stars. He discovered that

beyond the stars of the sixth magnitude you will behold through the telescope a host of other stars, which escape the unassisted sight, so numerous as to be almost beyond belief, for you may see more than six other differences of magnitude, and the largest of these, which I may call stars of the seventh magnitude, or of the first magnitude of invisible stars, appear with the aid of the telescope larger and brighter than stars of the second magnitude seen with the unassisted sight. But in order that you may see one or two proofs of the inconceivable manner in which they are crowded together, I have determined to make out a case against two star-clusters, that from these as a specimen you may decide about the rest.

As my first example, I had determined to depict the entire constellation of Orion, but I was overwhelmed by the vast quantity of stars and by want of time, and so I have deferred attempting this to another occasion, for there are adjacent

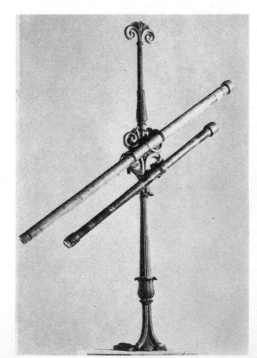

87 Two of the first telescopes
made in Italy, probably by
Galileo himself.

119

to or scattered among the old stars more than five hundred new stars within the limits of one or two degrees. For this reason I have selected the three stars in Orion's Belt, and the six in his Sword, which have been long well-known groups, and I have added eighty other stars recently discovered in their vicinity. . . . As a second example, I have depicted the six stars of the constellation Taurus, called the Pleiades (I say six intentionally, since the seventh is scarcely ever visible), a group of stars which is enclosed in the heavens within very narrow precincts. Near these there lie more than forty others invisible to the naked eye, no one of which is more than half a degree off any of the aforesaid six. . . .

. . . The next object which I have observed is the essence or substance of the Milky Way. By the aid of a telescope anyone may behold this in a manner which so distinctly appeals to the senses that all the disputes which have tormented philosophers through so many ages are exploded at once by the irrefragable evidence of our eyes, and we are freed from wordy disputes upon the subject, for the galaxy is nothing else but a mass of innumerable stars planted together in clusters. Upon whatever part of it you direct the telescope straightway a vast crowd of stars presents itself to view; many of them are tolerably large and extremely bright, but the number of small ones is quite beyond determination.

Galileo concluded with

the matter which seems to me to deserve to be considered the most important in this work, namely, that I should disclose and publish to the world the occasion of discovering and observing four planets, never seen from the very beginning of the world up to our own times. . . . On the 7th day of January in the present year, 1610, in the first hour of the following night, when I was viewing the constellations of the heavens through a telescope, the planet Jupiter presented itself to my view, and as I had prepared for myself a very excellent instrument, I noticed a circumstance which I had

88 Page from Galileo's notebook recording his observations of Jupiter and its satellites.

never been able to notice before, owing to want of power in my other telescope, namely that three little stars, small but very bright, were near the planet.

By careful observation of the changing position of these three 'stars' relative to Jupiter, Galileo had concluded by 11 January,

that there are three stars in the heavens moving about Jupiter, as Venus and Mercury round the sun; which at length was established as clear as daylight by numerous other subsequent observations. These observations also established that there are not only three but four erratic sidereal bodies performing their revolutions round Jupiter.

This discovery of the four largest satellites of Jupiter was, as Galileo noted, a

splendid argument to remove the scruples of those who can tolerate the revolution of the planets round the sun in the Copernican system, yet are so disturbed by the motion of one moon about the earth, while both accomplish an orbit of a year's length about the sun, that they consider that this theory of the universe must be upset as impossible: for now we have not one planet only revolving about another, while both traverse a vast orbit about the sun, but our sense of sight presents to us four satellites circling about Jupiter, like the moon about the earth, while the whole system travels over a mighty orbit about the sun in the space of twelve years.

Galileo's telescopic observations, by revealing whole new aspects of the universe, struck a series of further blows against Aristotelian cosmology. The revelation that the moon was composed of material much like that of the earth shattered the Aristotelian idea that the heavenly bodies were all perfect and incorruptible objects composed of an unchanging 'quintessence'; the discovery of a host of new stars, invisible to the naked eye, drew attention to questions about the size and possible infinity of the universe which had been in vogue since Copernicus's work; and the proof that Jupiter had four satellites, showing that the earth was not the only body in the heavens to have a moon circling round it, made Copernican ideas seem more plausible to many, as Galileo himself was quick to point out.

Galileo's astronomical work thus gave a considerable impetus to the rejection of Aristotelian ideas about the universe and to the acceptance by educated men of the Copernican system which Galileo himself later defended so effectively in his *Dialogue on the Two Chief Systems of the World*. That work, like the *Two New Sciences*, was written in Italian, instead of the Latin which was still traditional for scholarly works, and this use of the vernacular, together with the vigour and clarity of his prose, ensured that Galileo's ideas received widespread publicity. Above all in his methodology Galileo revealed himself as the first unquestionably 'modern' scientist – a man whose approach to scientific experiments was not essentially different from that of his twentieth-century successors. Of course, despite all his great contributions to scientific advance, Galileo's work had its weaknesses – his belief, for instance, in the circularity of both inertial motion and of the planets' orbits. Most important of all, he tended to regard the problems of mechanics and astronomy as separate issues. He did not see their fundamental interconnections and did not, therefore, try to unite them in a comprehensive system. That was left as the great task of Isaac Newton.

NEWTON

Newton was born in the little village of Woolsthorpe in Lincolnshire on Christmas Day 1642. He was a tiny, weakly infant, so small at birth that his mother later said that he could have been put into a quart pot. He showed no special brilliance at school, but after he went up to Trinity College, Cambridge, in 1661 he came under the influence of Professor Isaac Barrow, a notable mathematician. In the autumn of 1665 fear of the plague led to the closing of the university and Newton returned to Woolsthorpe for eighteen months until it reopened in the

spring of 1667. That comparatively short space of time was one of the two great creative periods of his scientific life. Indeed, looking back on these years when he was an old man, Newton regarded them as the most significant time of all. 'In those days I was in the prime of my life for invention and minded mathematics and philosophy more than at any time since.'

All Newton's subsequent work was, in fact, very largely a development of the discoveries he made at that time. During these years he invented the calculus, did fundamentally important researches into the composition of light, and, above all, worked out an early version of the law of gravitation as it applied to the earth and the moon. He published nothing, but went back to Cambridge, and in 1669 Barrow, who recognized his student's stupendous genius, resigned the Lucasian Chair of Mathematics in his favour. Newton retained the Chair until 1701 and, as it carried light duties and a competent income, it left him plenty of time to devote to his own interests. He returned to the work on light which he had started in 1666, and in 1672 and 1675 published papers on the subject. Much later he collected all his views on light in a book which appeared in 1704 entitled *Opticks*. This is a scientific masterpiece, written in prose of limpid clarity and containing details of experimental investigations of the phenomena of light which are models of scientific methodology. Newton's theories of light provoked considerable controversy at the time, notably with Robert Hooke, a fellow member of the Royal Society, who argued that light consisted of a series of waves transmitted through the ether which pervaded space. Newton, on the other hand, maintained that light was made up of particles or corpuscles which emanated from luminous bodies and gave rise to waves as they passed through the ether. Newton thus advanced a combination of wave and corpuscular theories, although he did lay emphasis on the corpuscular aspects. His ideas went out of fashion in the nineteenth century, when wave theories of light were dominant, but they have returned to favour in the twentieth century, which has seen a fusion of wave and particle theories.

The *Opticks* was certainly a work of genius, but Newton's claim to immortal fame rightly rests on his *Philosophiae Naturalis Principia Mathematica* (*Mathematical Principles of Natural Philosophy*), universally known as the *Principia*, which was published in 1687. In the following years he went through periods of acute depression which culminated in a nervous breakdown in 1693. During that time he showed little interest in science and made considerable efforts to obtain an administrative post. He was not immediately successful, but eventually, in 1696, thanks to the good offices of his friend Charles Montague,

90 Diagram from Newton's *Opticks*: a beam of white light is passed through a series of prisms and lenses that split it into a coloured spectrum, recombine it into a single beam, then divide it once more.

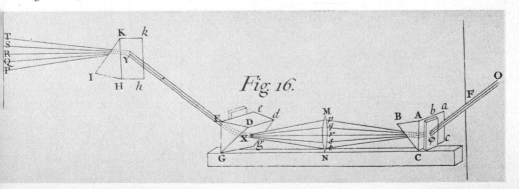

then Chancellor of the Exchequer, he was appointed Warden of the Royal Mint. As such, he was immediately under the chief officer, the Master of the Mint, and succeeded to the Mastership in 1699, retaining the office until his death. On being appointed Warden he moved from Cambridge to London, where he became President of the Royal Society in 1703 and held undisputed sway as the dictator of English science. He was knighted in 1705, and on his death in 1727, at the age of eighty-four, was buried with great ceremony in Westminster Abbey, the only English scientist ever to be honoured in this way.

Newton's was a very complex character. As a scientist he was a devoted experimenter and mathematical analyst who continually applied in all his work the modern methodology which had first been used by Galileo. But science was only one of his interests. He spent a great deal of time in theological and alchemical studies, leaving nearly two million words in manuscript on these subjects. His writings on alchemy included notes about the elixir of life and the transmutation of base into precious metals. It is such interests that have enabled one authority, the late Lord Keynes, to say that Newton's deepest instincts were 'occult' and 'esoteric' and to christen him 'the last of the magicians'. That judgment is clearly only a partial reflection of Newton's character, but it does serve to remind us that those modern authors who have seen him merely as the greatest and most rational of scientists are also wide of the mark.

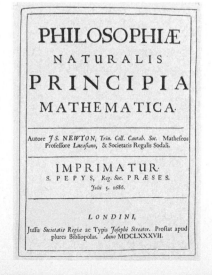

91   Title-page of Newton's *Principia*. The book carries the imprimatur of Samuel Pepys, then President of the Royal Society.

DE

MOTU CORPORUM

Liber PRIMUS

SECT I.

De Methodo Rationum primarum & ultimarum, cujus ope sequentia demonstrantur.

LEMMA I.

Quantitates, ut & quantitatum rationes, quæ ad æqualitatem dato tempore constanter tendunt & eo pacto propius ad invicem accedere possunt quam pro data quavis differentia, fiunt ultimo æquales.

Si negas, sit earum ultima differentia *D*. Ergo nequeunt propius ad æqualitatem accedere quam pro data differentia *D*: contra hypothesin.

Lem-

27

Lemma II.

Si in figura quavis Aa cE rectis Aa, AE, & curva AcE comprehensa, inscribantur parallelogramma quotcunq; Ab, Bc, Cd, &c. sub basibus AB, BC, CD, &c. æqualibus, & lateribus Bb, Cc, Dd, &c. figuræ lateri Aa parallelis conten-ta; & compleantur parallelogramma aKbl, bLcm, cMdn, &c. Dein horum parallelogrammorum latitudo minuatur, & numerus augeatur in infinitum: dico quod ultimæ rationes, quas habent ad se invicem figura inscripta AKbLcMdD, circumscripta AalbmcndoE, & curvilinea AabcdE, sunt rationes æquales.

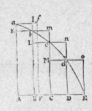

Nam figuræ inscriptæ & circumscriptæ differentia est summa parallelogrammorum $Kl+Lm+Mn+Do$, hoc est (ob æquales omnium bases) rectangulum sub unius basi Kb & altitudinum summa Aa, id est rectangulum $ABla$. Sed hoc rectangulum, eo quod latitudo ejus AB in infinitum minuitur, fit minus quovis dato. Ergo, per Lemma I, figura inscripta & circumscripta & multo magis figura curvilinea intermedia fiunt ultimo æquales. Q. E. D.

Lemma III.

Eædem rationes ultimæ sunt etiam æqualitatis, ubi parallelogrammorum latitudines AD, BC, CD, &c. sunt inæquales, & omnes minuuntur in infinitum.

Sit enim AF æqualis latitudini maximæ, & compleatur parallelogrammum $FAaf$. Hoc erit majus quam differentia figuræ inscriptæ & figuræ circumscriptæ, at latitudine sua AF

E 2   in

---

92   Two pages from the *Principia*, discussing the momentum of bodies.

He was certainly an aloof man, afraid of close contacts with his fellows and inordinately suspicious of their attitude towards himself and his work. He always steered clear of women, and his lack of any normal sex life has made him an obvious target for psychiatric study and has resulted in a recent massive Freudian biography.

Clearly it is impossible to reach any final judgment about the inner life of such a man, but one thing is certain: he had a stupendous intellect. An admirer is said to have asked him how he made his discoveries. 'By always thinking unto them', he replied, and Keynes has argued that the secret of his scientific genius did indeed lie in unexampled powers of concentration – 'the ability to hold a problem in his mind for hours and days

127

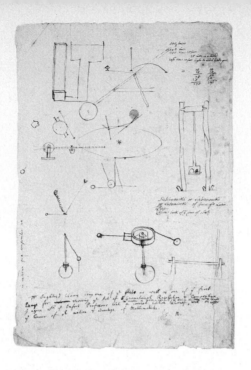

93 Drawings of apparatus by Newton, probably relating to pendulum experiments.

and weeks until it surrendered . . . its secret.' Newton in fact was a 'consecrated solitary, pursuing his studies by intense introspection with a mental endurance perhaps never equalled'.

It was that supreme intellectual ability which enabled him to produce the *Principia*, the work which inspired Alexander Pope to write his celebrated 'Epitaph intended for Sir Isaac Newton':

> *Nature and Nature's laws lay hid in night:*
> *God said, Let Newton be! and all was light.*

This fitting tribute must not, however, obscure the fact that the *Principia* was firmly founded on the work of Newton's predecessors and contemporaries. He owed, for example, considerable debts to Galileo, Kepler and Descartes, and to Christiaan Huygens, who had worked out the first satisfactory theory of centrifugal force. This, of course, is not to belittle in any way Newton's own achievement. It was he who conceived the universal application of the law of gravitation, which formed the central feature of the *Principia*, and he alone was capable of producing the detailed mathematical proofs which demonstrated the practical workings of the law. Encouraged by Edmund Halley, a distinguished fellow scientist, he began to write his great work in 1684, finishing it, as he himself said, 'in seventeen or eighteen months', in the spring of 1686. This was the second period of sustained scientific creativity in his life, and at this time he employed a secretary, who has left us the following account of his working habits.

> I never knew him to take any recreation or pastime either in riding out to take the air, walking, bowling, or any other exercise whatever, thinking all hours lost that was not spent in his studies, to which he kept so close that he seldom left his chamber except at term time, when he read in the schools. . . . He very rarely went to dine in the hall, except on the same public days, and then if he has not been minded would go very carelessly, with shoes down at heels, stockings untied, surplice on, and his head scarcely combed. At some seldom times when he deigned to dine in the hall, [he] would turn to the left hand and go out into the street. . . . When he found his mistake [he] would hastily turn back, and then sometimes instead of going into the hall, would return to his chamber again.

Usually, in fact, Newton had his meals sent up to his rooms – and then forgot to eat them!

The *Principia*, the result of this intense and sustained mental effort, is not an easy book to read – it has been suggested, probably rightly, that only three or four of Newton's European contemporaries were even capable of understanding it fully. The problems arise partly from the intrinsic difficulty of some of the material, but also from the mathematical machinery employed in the book. Although Newton used the calculus to obtain some of his results, he cast the final work in the forms of classical geometry, just at the time when such methods were going out of fashion. In the words of William Whewell, a famous nineteenth-century Cambridge scholar: 'Nobody since Newton has been able to use geometrical methods to the same extent for the like purpose, and as we read the *Principia* we feel as when we are in an ancient armoury where the weapons are of gigantic size; and as we look at them we marvel what manner of man he was who could use as a weapon what we can scarcely lift as a burden.'

The *Principia* consists of three books. Newton began with a number of definitions and axioms in which he made explicit for the first time the basic concepts of mechanics, such as force, momentum and mass, and stated his three laws of motion, which are the foundation of classical dynamics. The first law – the law of inertia – states that every body continues in its state of rest or of uniform motion in a straight line unless compelled to change that state by forces impressed upon it. The second law, which deals with acceleration, states that the change of motion of a body is proportional to the force acting upon it and takes place along the straight line along which the force acts. The third law, dealing with the equivalence of action and reaction, states that to every action there is always an equal and opposite reaction; or, the actions of two bodies upon each other are always equal and opposite.

In the first book of the *Principia* Newton discussed the motions of bodies in unresisting mediums and provided the entire mathematical theory on which celestial mechanics is based. This part of the work is wholly general, consisting of a number of geometric theorems on such subjects as the determination of orbits and the attraction exercised by spherical bodies. It was in this book that Newton demonstrated that, for purposes of calculation, the heavenly bodies could be treated as if their entire masses were concentrated at their central points – a taxing mathematical problem. In Book Two Newton considered the motions of bodies in resisting mediums, in other words, in fluids. This is in some ways a digression from the main argument, though near the end he showed conclusively that Descartes's idea that the universe was filled with a kind of celestial fluid

94 One of Newton's manuscripts, illustrating the diversity of his interests. The top portion shows calculations on motion in a resisting medium. The middle section deals with currency in Brandenburg and France. At the bottom is a theological note. ▶

$co \ \square\square\square$. $\frac{ao}{e} = 9F$. $GF = \frac{nno^2}{2e^3} + \frac{a}{amo}$. $CF = \sqrt{oo + aaoo} = \frac{no}{e}$. $\xi f = \frac{nnoo}{2e^9} + \sqrt{\frac{annob}{2e^5}} + oe$

$= \frac{nupp}{2e^3} - \frac{annp^3}{2e^5}$. $pp - \frac{ap^3}{ee} = \frac{ao}{2e} + \frac{ao^3}{ee}$. $oo. pp:: ee - ap. ee + ao.$

$o. p - o:: \varepsilon. \frac{ao}{2e}:: BC. \mathcal{G}F:: CF. kf = Cf - CF = 2FH = \frac{naoo}{2e^3}$. $\frac{naoo}{2e^3} = TH. \frac{nnoo}{2e^3} = T\mathcal{G}:: A. n$

Ergo $GH \perp FH$ et corpus non accelleratur.

$\frac{naoo}{2e^3} = Cf - CF$. $\frac{a}{n} = \frac{Cf - CF}{FG}$ ut resistentia. $\frac{a}{2ne}$ ut densitas.

$GF = \frac{CF^9}{2GD} = dato$. CF^9 ut $\overline{GD}$. CF ut $\overline{GD}^{\frac{1}{2}}$

Velocitas ut $GD^{\frac{1}{2}}$. Decrementum $\mathcal{G}$

$\mathcal{G}D$ ut $GD^{\frac{3}{2}}$. $4\mathcal{G}$ ut $OD. DG^{\frac{1}{2}}$, decrementum

velocitatis ut$\mathcal{P}$ decrementum $\square^{li}$ velocitati.

$\frac{Cf - CF}{2} = HF$. $\frac{CF^9}{2C3}::$ resist. Grav.

$\sqrt{fg} \times \overline{fd + dg} - \sqrt{FG} \times \overline{FD + GD}$. $FG:: \sqrt{fd + dg} - \sqrt{FD + DG}$. $2\sqrt{FG}:: \sqrt{dg} - \sqrt{DG}. \sqrt{2FG}$

$Ct = CFG$. $OB = a. BC = e. BD = o = Bd i. Bd = p.$

$\frac{ann}{2e^5} \times \frac{n}{2}$ ad $\frac{nt}{2e^8}$ ut $\frac{a}{2}$ ad $\frac{n}{2ee}$

$2\beta e_3 \sqrt{nt + ll oo}$ ad $\sqrt{\frac{t}{co}}$ $4RRo4$

Grav. Resist ::

NB. 1. The Electors of Brandenburg Saxony & the House of Lunenburg have agreed
to coin their moneys of equal value but the agreed though not of equal alloy.

2 that the French mark of eight ounces equals 7 ounces
French for gold & silver is toy ounce Troy of ye Excheqr & Tower

2 The ounce French 59 to 60 $26650 (24009$

$222. 164 :: 111. 82 :: 325. 186$

$222. 16\frac{1}{2} :: 178. 1\xi536.$

193.536

$8.1.7. | 25.$

$440.1 :: 165.75$

$6. 21.7\frac{1}{2} | 21.$

$27. 6. 74$

Martyr seems to have read $\overline{\varepsilon}$ or $\overline{\varepsilon}$. For ε or $\overline{\varepsilon}$ For
so Justin, in his Epistle to Diognetus, in saying $O\tilde{v} \chi \acute{\alpha} \rho\iota\nu$
$\dot{\alpha}\pi\acute{\epsilon}\sigma\tau\epsilon\iota\lambda\epsilon$ $\Lambda\acute{o}\gamma o\nu$, $\tilde{\iota}\nu\alpha$ $\kappa\acute{o}\sigma\mu\omega$ $\varphi\alpha\nu\tilde{\eta}$: $\ddot{o}\varsigma$, $\dot{\upsilon}\pi\grave{o}$ $\Lambda\alpha\tilde{\omega}$ $\dot{\alpha}\tau\iota\mu\alpha\theta\epsilon\grave{\iota}\varsigma$, [For wch
$\dot{\alpha}\pi o\sigma\tau\acute{o}\lambda\omega\nu$ ... $\dot{\upsilon}\pi\grave{o}$ $\check{\epsilon}\theta\nu\tilde{\omega}\nu$ $\dot{\epsilon}\pi\iota\sigma\tau\epsilon\acute{\upsilon}\theta\eta]$ that he might appear to the world: who, being
despised by the people, & preached by the Apostles, was believed on by
the Gentiles. Had he interpreted not $\Theta\epsilon\acute{o}\varsigma$ God but
$\ddot{o}$ or $\ddot{o}\varsigma$. Had $\Theta\epsilon\acute{o}\varsigma$ been in his text it would have needed no interpretation.

For that Photinus held the Word to be the $\lambda\acute{o}\gamma o\varsigma$ emitted outwardly
as it were by speaking before the World began is signified also by Ambrose
in these words. Et Deus erat Verbum. Non ergo in prolatione sermonis hoc
Verbum est, sed in illa caelestis designatione virtutis, ut confutetur Photinus.
Ambr. l. 1 de fide, c. 4. And Augustin tells us that Photinus

through which the stars and planets swirled at the centre of vortices was impossible on strictly mechanical grounds.

Book Three is the culmination of the work. In it Newton systematically applied his mechanical theories to the problems of astronomy. He himself stated at the beginning: 'In the preceding books I have laid down the principles of philosophy [science]. . . . These principles are the laws and conditions of certain motions, and powers or forces, which chiefly have respect to philosophy. . . . It remains that from the same principles I now demonstrate the frame of the System of the World.' These were grandiose words, but he justified them by demonstrating that the mechanical ideas which he set out in Book One, when applied to the velocities, masses and distances of the known bodies of the solar system, accurately confirmed all the phenomena already established about these bodies. The vital feature in the whole argument was Newton's law of gravitation, the idea that every body, indeed every particle of matter in the universe, exerts over every other body or particle an attractive force proportional to the product of their masses and inversely proportional to the square of the distance between them.

Newton's theory of gravitation was a universal law which explained the whole physical structure of the universe and unified the phenomena of earth and heavens which had been so entirely separate in the Aristotelian synthesis. It also ended the breach between physical reality and mathematical astronomy which had characterized medieval science. Copernicus had begun the attack by his challenge to the Ptolemaic system, but it was only completed in the *Principia*, which pictured a universe where physics and astronomy explained and complemented rather than contradicted each other. The sun and its attendant planets formed a self-contained system whose motions were determined by the mutual interaction of its components. The stars were scattered through an unimaginable vastness of space, far beyond the reaches of the solar system. They too, of course, both exercised and were subject to the force of gravity, but they were much too distant to make their influence felt on earth. That Newtonian conception of the universe is a recognizable one: it is the beginning of our modern picture.

It would be quite wrong to suppose that Newton's ideas immediately carried all before them in the scientific world. It is true that they were almost immediately accepted in England – as much perhaps for reasons of national pride as absolute conviction – but they aroused strong resistance on the Continent, where such scientists as Leibniz and Huygens expressed grave reservations. Opposition was particularly strong in France where Descartes's ideas had just conquered scientific thought, and where Bernard de Fontenelle, secretary of the French Academy, remained a convinced follower of Descartes. By the time of his death in 1757, however, he was fighting a losing battle, for, partly through the influence of Voltaire, Newtonian ideas had almost completely triumphed in France; and from there they spread to much of the rest of Europe.

The general acceptance of Newton's mechanistic ideas led, in the eighteenth century, to the detailed application of his theories to astronomical problems and to the use of some of his essential principles in other branches of science. In the later part of the century the French mathematicians Joseph Lagrange and Pierre de Laplace carried out a detailed survey of the Newtonian theory of the solar system, showing that it was a self-regulating mechanism. Newton, building on Galileo's ideas, had taken one phenomenon – motion – exhibited by bodies of all kinds, irrespective of their other characteristics (such as weight, size and colour) and had selected concepts (such as mass and force) in terms of which these motions could be discussed. In the eighteenth century scientists applied this essential Galilean-Newtonian methodology to the study of such subjects as heat, light, magnetism and electricity, also choosing appropriate concepts in which they could be discussed. The formulation of general laws based on these concepts was essentially the work of the nineteenth century.

The belief that Newton had explained once and for all the fundamental mechanism of the universe, a belief prevalent from the eighteenth century right up to the twentieth, had probably a good deal to do with the fundamental optimism of the eighteenth and nineteenth centuries – the belief that man, by the exercise of his reason, could explain all natural phenomena and improve his own condition on earth. That modern belief in continuous progress which originated in the seventeenth century only became really widespread in the eighteenth, for example in the work of the *philosophes*.

Newtonian science also gave rise to determinism, a fact which had important religious implications. Newton himself and the first generation of his followers believed in the existence of God as a dynamic, ever-present force in the universe, who ran it actively according to His own laws. The actual mechanism of the Newtonian universe could, however, be easily reconciled with a God who acted merely as a 'first mover' and was not called upon to do anything further after His original creation of the universe, natural laws taking over all the subsequent operations. This deterministic view, which aided the development of deism, was expressed by Laplace, who, when asked by Napoleon about the position of God in his system, replied firmly, 'I do not need that hypothesis.'

MEDICAL ADVANCES

The work of the scientific revolution in biology, notably in the medical sciences, was much less fundamentally important than its work in physics and astronomy, but it did lay the foundations for subsequent improvements in anatomy, physiology and medical practice, and must, therefore, receive at least brief attention in any discussion of the scientific changes of the period.

The first notable figure in sixteenth-century medical history is the Swiss-German physician Philippus von Hohenheim, who called himself Paracelsus. He was born near Zürich in 1493 and must have picked up some knowledge of medicine from his father, who was the local physician. During his teens and early twenties he travelled widely in both Europe and the Middle East, working as an army surgeon. He also visited a number of

FAMOSO·DOCTOR PARESELSVS.

95 Philippus von Hohenheim, who styled himself Paracelsus to indicate that he was greater than Celsus (a famous Roman writer on medicine who lived during the first century A D).

continental universities and may have taken a medical doctorate at Ferrara. He was certainly an extremely quarrelsome and conceited man and a well known address to fellow physicians indicates why he always fell out with his contemporaries.

> I am *monarcha medicorum*, and I can prove to you what you cannot prove. . . . It was not the constellations that made me a physician; God made me. . . . I need not don a coat of mail or a buckler against you, for you are not learned enough to refute even one word of mine. . . . Let me tell you this: every little hair on my neck knows more than you and all your scribes . . ., and my beard has more experience than all your high colleges.

His contempt for his fellows extended to the great medical authorities of the past, as he made clear when he described his appointment in 1527 as city physician and professor of medicine at Basle.

> Being invited by an ample salary of the rulers of Basle, for two hours in each day [I] do publicly interpret the books

both of practical and theoretical medicine, physics and surgery, whereof I myself am author, with the greatest diligence, and to the great profit of my hearers. I have not patched up these books, after the fashion of others, from Hippocrates, Galen or anyone else, but by experience the great teacher, and by labour, have I composed them. Accordingly, if I wish to prove anything, experiment and reason for me take the place of authorities.

He drove the point home by burning the *Canon* of the great medieval Moslem physician Avicenna at the traditional St John's Day bonfire in the city. This action, together with his other violent prejudices, was too much for the townsmen, and he was forced to leave Basle. For the rest of his life he wandered from one town to another, never staying anywhere for very long. In 1541 he died, at the age of forty-eight.

Paracelsus's medical ideas derived from his general views about the universe. He believed that the creation had been a kind of 'divine chemical separation' when the four Aristotelian elements, earth, water, air and fire, together with three 'principles', sulphur, mercury and salt were created. He thought that human beings reproduced in their bodies, on a tiny scale, the chemical reactions of the universe as a whole, and rejected the Galenic idea that disease was caused by an imbalance of four humours which created a disturbance of the whole body. He taught instead that diseases were localized in particular organs, were due to chemical influences, and could be effectively treated by chemical remedies. Indeed, according to Walter Pagel, his most notable recent biographer, 'he was actually the first [man] to teach that there are different diseases which can be classified and that each disease is a peculiar reality.' Although he was by no means the first to use chemically prepared medicines in his treatments, he placed such emphasis on them in opposition to the traditional Galenic herbal remedies, that his name was soon irrevocably associated with them, and he was hailed as the founder of medical chemistry (iatro-chemistry).

Paracelsus laid considerable stress on the importance of determining the correct dosages to be given to patients. This was a new emphasis in medicine and in Paracelsus's case an essential precaution as well, as he believed that small doses of poisonous substances were among the most effective drugs. He thought, in fact, that like cured like, that the proper dosage of a poison which had caused a disease would also cure it. These homoeopathic ideas, which followed ancient folk traditions, were also a challenge to established authority, as Galen had believed that 'contraries' cured, that, for example, a medicine with an excess of the 'cold' quality would restore the balance of a humoral system that had become overheated.

By the early seventeenth century Paracelsus's ideas had spread widely throughout Europe. England was influenced by his thought from the 1570s onwards, but the disputes between English Paracelsians and their more conservative colleagues seem to have been a good deal less bitter than those on the Continent, where there were violent conflicts between the followers of Galen and Paracelsus.

Paracelsus's positive contributions to medicine were mainly in the diagnosis and treatment of disease. The important work of his junior, Andreas Vesalius, lay, in contrast, in the field of anatomy. Vesalius was born at Brussels in 1514 and started university studies, in the humanities, at Louvain. In 1533 he began informal medical training at Paris, where he was profoundly influenced by the great current interest in the works of Galen. That had been stimulated by the recent discovery of Galen's book, *On Anatomical Procedures*, and Vesalius, although he later improved on much of Galen's work, owed considerable debts to him. Galen taught him, above all, to regard anatomy as a subject for practical research as opposed to the merely theoretical exposition which was the rule in the Middle Ages. In 1536 he left Paris for Padua, where he became a doctor of medicine in the following year and was immediately appointed professor of surgery. He spent the next six years teaching and preparing his lectures for publication. They came out in 1543, in his book *De Humani Corporis Fabrica* (*On the Fabric of the Human Body*). After its publication he gave up teaching and spent most of the rest of his life as physician, first of all to the Emperor Charles V and then to Philip II of Spain. At the very end of his days he planned to return to Padua but died in 1564 before he could do so.

Vesalius was a great success as a teacher. The novelty of his method lay in the fact that at his lectures he dissected bodies himself, instead of adhering to the traditional practice of leaving the work to an assistant while he expounded a Galenic text from a raised chair. The title-page of the *De Fabrica* made this point by showing Vesalius demonstrating directly from a corpse. The *De Fabrica*, the end product of all this practical work, is a magnificent anatomical treatise in which the text and plates complement each other. In it Vesalius carefully examined both the general structure and each of the organs of the body. During his own researches he had become increasingly aware of the fact that Galen's descriptions, based largely on the dissections of animals, were often at fault. He did not correct by any means all of these errors, but the text of the work as a whole undoubtedly represented a substantial advance in anatomical knowledge. Much of its value depended on its superb illustrations of the human skeleton, muscles and organs, far superior to any previous detailed pictures of the body. The unknown artists who drew them must have been men of the highest abilities and in their final form, as they are reproduced in the *De Fabrica*, the illustrations are both a tribute to the art of printing in the early sixteenth century and a remarkable example of the importance which printing techniques had acquired by that time in the dissemination of scientific knowledge.

Although Vesalius overthrew some of Galen's ideas about the structure of the human body, he accepted in its essentials the latter's account of its functioning or physiology, based on the idea that the blood ebbed and flowed upwards and downwards in the veins and arteries. This theory was not rejected until the work of William Harvey in the early seventeenth century.

138

96   Vesalius dissecting a corpse: title-page of *De Fabrica*.   ▶

ANDREAE VESALII
BRVXELLENSIS, SCHOLAE
medicorum Patauinæ professoris, de
Humani corporis fabrica
Libri septem.

CVM CAESAREAE
M.aiest. Galliarum Regis, ac Senatus Veneti gratia & priuilegio, ut in diplomatis eorundem continetur.

97, 98, 99 Left, and opposite: successive
layers of muscle revealed in three splendid
illustrations from Vesalius's *De Fabrica*.

Harvey was born in 1578, the son of a Kentish yeoman farmer.
He was educated at the King's School, Canterbury, at Caius
College, Cambridge, and then from 1600 at the university of
Padua, where he studied under the distinguished anatomist
Hieronymus Fabricius and graduated doctor of medicine in
1602. He immediately returned to England and soon established
a reputation in London, where he was appointed physician to
St Bartholomew's Hospital in 1609, later becoming physician
extraordinary to both James I and Charles I. He combined his
medical practice with research and in 1628 published the work
on which his fame rests, *De Motu Cordis et Sanguinis* (*On the
Motion of the Heart and Blood*). During the remainder of his life

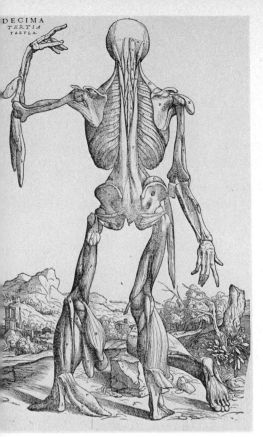
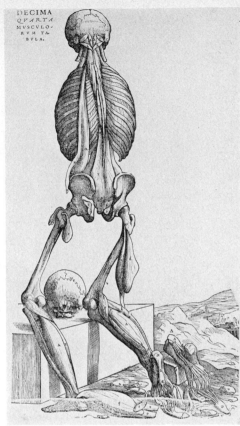

he produced further important researches, chiefly in the field
of embryology. He died in 1657, at the age of seventy-nine.

In the sixteenth century, long before Harvey published his
theory of the circulation of the blood, evidence was accumulat-
ing about the inadequacies of Galenic physiology. For example,
Galen's ideas required blood to travel right through the central
wall of the heart. Vesalius himself investigated this central wall
and could find no passages. 'None of these pits', he stated,
'penetrate from the right ventricle to the left.' Instead of con-
cluding, however, that the blood could not pass that way, he
found himself 'compelled to marvel at the activity of the creator
of things, in that blood should sweat from right ventricle to the

141

100 (Overleaf) John Banister, a prominent English anatomist, lecturing in London
in 1581.

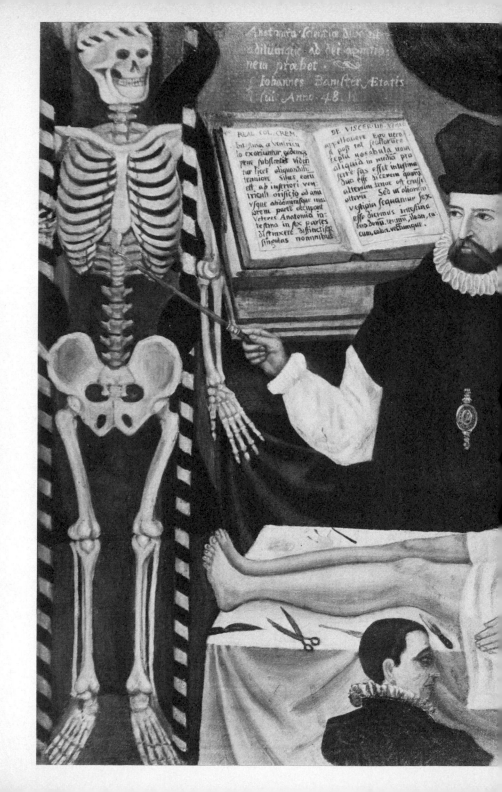

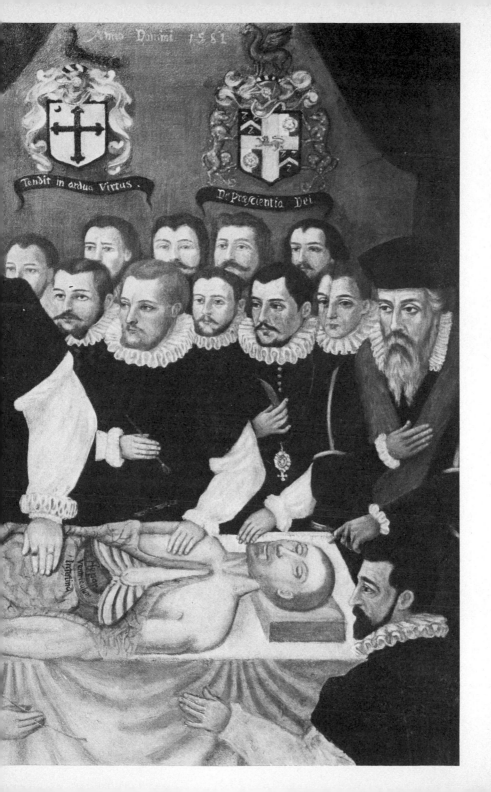

101 The famous anatomy theatre at Padua University, where Vesalius and William Harvey were students, appears on the left of this view of the university.

left through passages escaping the sight.' This remarkable statement by the normally independent-minded Vesalius bears witness to the tremendously strong hold which the basic ideas of Galenic physiology still exercised.

In 1553 the Spaniard Miguel Serveto offered an alternative to the theory of the passage of blood through the central wall of the heart, when he correctly described the 'lesser circulation' of the blood, that is to say its passage from the left ventricle to the right auricle of the heart by way of the lungs. This idea was accepted by some authorities, but it does not seem to have suggested to anyone the possibility that all the blood might circulate. Further important information was provided by Harvey's own teacher Fabricius, who in 1603 published a notable work *De Venarum Ostiolis* (*On the Valves in the Veins*). However, he

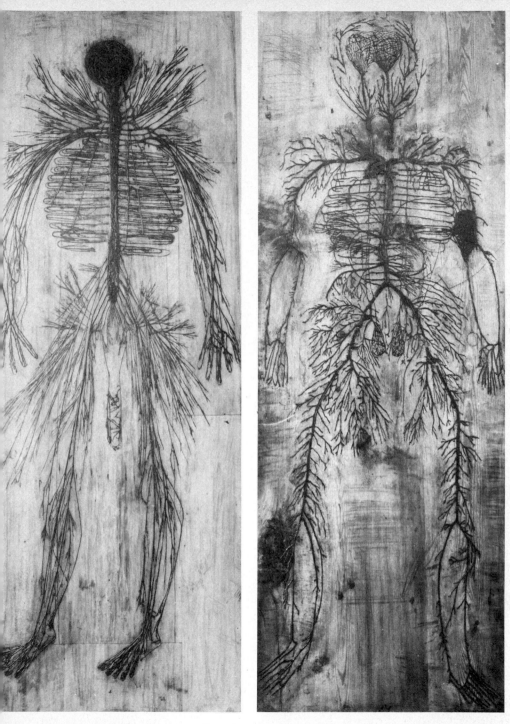

102, 103   Seventeenth-century dissections, showing the nervous system and, right, the arterial system.

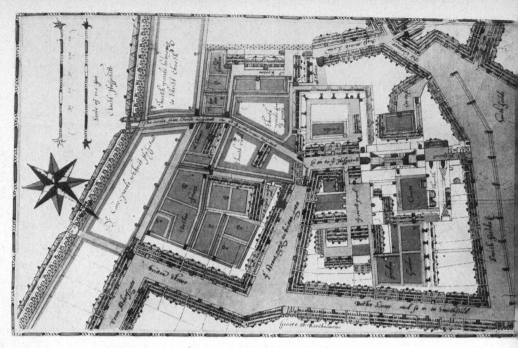

104 St Bartholomew's Hospital, London, a centre for the advancement of medical knowledge in seventeenth-century England.

had begun to discuss these valves in his lectures in the 1570s, so Harvey had learnt about them during his years at Padua. The valves permit venous blood to flow in one direction only – towards the heart – and seem to have given Harvey himself a vital clue when he was framing his ideas about the circulation of the blood. Fabricius, however, completely misinterpreted the function of these valves, arguing that it was to delay the passage of the blood away from the heart, thus preventing all the blood from collecting in the feet or hands.

Sixteenth-century medical experts, therefore, did not break away from basic Galenic physiology. That daring step was accomplished by Harvey in the early years of the seventeenth century. Harvey had been making meticulous examinations of the insides of many different kinds of living animals for years

Gulielmus
(Magnus ille)
Harveus

105 Portrait of William Harvey, whose theory of the circulation of the blood was
based on precise observation and experiment.

before he published his book. As a result the *De Motu* is a model of inductive reasoning, based on careful observations and experiments. Harvey showed, among other things, that the dynamical starting-point of the blood is the heart and not the liver, as Galen had supposed; that there are no pores in the central wall of the heart; that the same blood flows in both arteries and veins; and that in its passage through the body the blood makes a complete 'circulation'. Harvey's decisive argument was a quantitative one. He calculated that the amount of blood pumped by the heart into the arteries in half an hour was greater than the total amount of blood in the body. It was impossible to account for this except on the basis of his circulation theory. Thus, only nineteen years after Kepler had rejected the idea that the heavenly bodies moved in circles, Harvey introduced the conception of circular movement within the human body. It was a complete reversal of traditional Aristotelian teaching.

There was one missing link in the chain of Harvey's evidence – proof that the body's blood passed from the finest arteries to the finest veins. As we have seen, these capillaries were not identified till 1661, and it was only about then that Harvey's ideas began to win general acceptance. He had, in fact, dealt the death blow to Galenic physiology. As no substantial physiological advances could possibly be made while such mistaken views held the field, Harvey's discovery of the circulation of the blood was a necessary basis for the creation of modern physiology.

These medical and biological developments made, however, a much less important contribution to the scientific revolution than the great discoveries in mechanics and astronomy. They did, it is true, give men a much more accurate knowledge of both the anatomy and physiology of their bodies, but this did not lead to major changes in outlook in any way equivalent to those initiated by the advances in the physical sciences. Indeed, improvements in biological knowledge only began to cause fundamental alterations in men's concepts about their

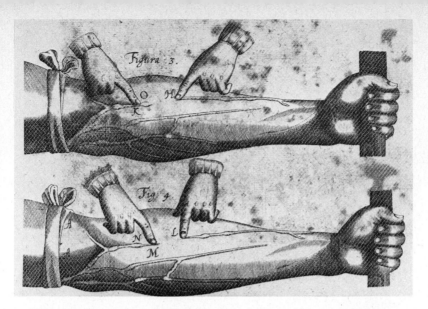

106 Diagram from Harvey's *De Motu* (1628), showing valves in the veins.

own physical evolution with the work of Charles Darwin in the nineteenth century.

DEVELOPMENTS IN CHEMISTRY
The advances of the seventeenth century in biology, physics and astronomy were the most significant developments of the time, but important work on chemistry was done by Robert Boyle. Boyle was born in 1627, the seventh son of the first Earl of Cork. His father's enormous wealth enabled him to devote himself from his early years to the scientific studies which fascinated him, and he was one of the leading influences behind the foundation of the Royal Society. From 1668 until his death in 1691 he lived in London at the home of his sister Lady Ranelagh, where, as a contemporary wrote, he had 'a noble laboratory and several servants to look to it'.

In the middle of the seventeenth century chemical ideas were still primitive, being usually based on the four elements of Aristotle or the three principles of Paracelsus. In his best-known work, *The Sceptical Chemist*, published in 1661, Boyle effectively

149

107 Robert Boyle, who
played a highly influential
role in the foundation of the
Royal Society: from a portrait
after John Kerseboom.

demolished these theories, but then wrongly questioned whether
chemical elements existed at all.

> I do not see [he wrote], why we must needs believe that there
> are any primogeneal and simple bodies, of which, as of pre-
> existent elements, nature is obliged to compound all others.
> Nor do I see why we may not conceive that she may produce
> the bodies . . . by variously altering and contriving their
> minute parts, without resolving the matter into any such
> simple or homogeneous substances as are pretended.

As this passage suggests, Boyle believed that the differences
between substances could be explained in terms of the different
arrangements of the ultimate particles of primary matter of
which they were all composed. He set these 'corpuscular' views
of matter out in detail in his *Origin of Forms and Qualities*, published in 1666.

Despite his rejection of the idea of chemical elements, Boyle
did a great deal of valuable scientific work. For example, he

made important classifications of a great range of different substances, and developed the famous law which bears his name and states that the volume of a gas varies inversely with the pressure exercised upon it. When all is said and done, however, there could be no revolution in chemistry without a proper understanding of the structure of matter, and this involved the recognition of elements. That is why Antoine Lavoisier, born in 1743 and executed by the French revolutionaries in 1794, who established once and for all the conception of chemical elements, is rightly regarded as having founded modern chemistry, with the publication of his classic textbook *La Traité Élémentaire de la Chimie (Elementary Treatise on Chemistry)* in 1789.

The discoveries of the sixteenth and seventeenth centuries did not, therefore, bring changes of equal significance throughout the whole field of science. There had been no revolution in chemistry by 1700, but the dramatic breakthroughs which had occurred by that date in astronomy, mechanics and mathematics had already had significant effects on the general ideas of educated Europeans.

108 The elaborate chemical laboratory of Ambrose Hanckwitz, who at one time worked as Boyle's assistant.

GLORIA SVMMA
NAM SVAIPSIVS SOLA

109 Literature was not the only art to be influenced by speculations on the nature
of the physical universe. This tapestry, with its astronomical motif, was com-
missioned for the Escorial, the new palace of Philip II of Spain, started in 1563.

IV THE SCIENTIFIC REVOLUTION: THE EFFECTS ON SOCIETY

THE GEOMETRICAL METHOD

The great French thinker Descartes, while still a boy, acquired a love of mathematics, because, as he put it, 'of the certainty of its proofs and the evidence of its reasonings'. That liking remained with him for the rest of his life, and he was the founder of *l'esprit géométrique*, which stressed clarity and calculation as a key to understanding. Bernard de Fontenelle gave the seal of his approval to such ideas:

> The geometrical method is not so rigidly confined to geometry itself that it cannot be applied to other branches of knowledge as well. A work on politics, on morals, a piece of criticism, even a manual on the art of public speaking would, other things being equal, be all the better for having been written by a geometrician. The order, the clarity, the precision and the accuracy which have distinguished the worthier kind of books for some time past now, may well have been due to the geometrical method which has been continuously gaining ground, and which somehow or other has an effect on people who are quite innocent of geometry. It sometimes happens that a great thinker gives the keynote to the whole of his century. He to whom the distinction of endowing us with a new method of reasoning may most justly be awarded [Descartes] was himself an accomplished geometrician.

These words were written at the end of the seventeenth century and there is no doubt that by 1700 the 'geometric spirit' and the general scientific attitudes of the time had begun to have significant effects on many fields of thought and human endeavour, for example on philosophy, literature and religion.

110 Bernard de Fontenelle (1657–1757), the greatest popularizer of new scientific ideas.

The word 'philosophy' was used in the seventeenth century to mean what is now called science, but here we are concerned with the term as it is understood in our own time: the study of the ultimate problems which are raised in man's search for truth about himself and the universe in which he lives. In the Middle Ages philosophers looked to authorities external to their own philosophical reasonings to confirm their ideas, notably to the Church and to the works of the ancient Greeks. Seventeenth-century thinkers rejected such limitations on their speculations. It is true that the Reformation, with its stress on faith as opposed to reason, had little direct influence on philosophical thought, but by flouting the teachings of the Church in purely religious matters the reformers weakened its authority in other fields as well. The scientific movement of the time, on the other hand, had direct and significant effects on philosophical developments. The work of such men as Copernicus, Galileo and Newton presented changed views of the universe which led philosophers to speculate anew about it and about man's place in it. Above all, the mathematical advances of the period had a profound effect on the leading seventeenth-century philosophers, all of whom, with the exception of John Locke, were themselves mathematicians of distinction.

The mathematical spirit so strongly apparent in Descartes's most famous philosophical work, the *Discours de la méthode,*

111   A diagram from Descartes' *Principia Philosophiae* (1644), showing the vortices, or whirlpools, which, he believed, carried the heavenly bodies through space. ▶

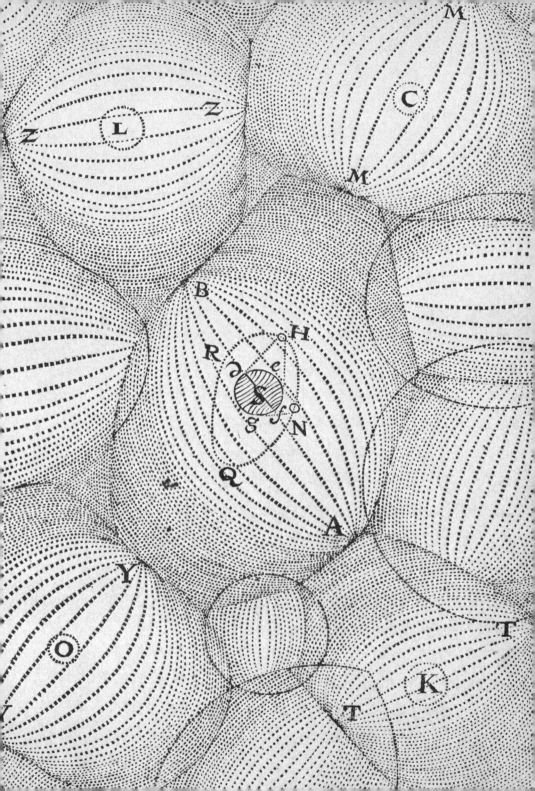

112 The French mathematician René Descartes, one of the most influential philosophers of the seventeenth century; his right foot rests on the works of Aristotle.

can also be clearly seen in the philosophies of the Dutchman Benedict de Spinoza and the German thinker Gottfried Leibniz. Spinoza agreed completely with Descartes's methods and set out his own philosophical conclusions, in the form of a highly deductive system, in his *Ethics*, which was published in 1677, while Leibniz took account of the works of both natural scientists and mathematicians in his philosophic thought. Leibniz, Spinoza and Descartes each attempted to construct an all-embracing philosophical system, but John Locke, the other outstanding philosopher of the period, abandoned such ambitions. He approached the problems of philosophy in a quite different way from Descartes, using what he called a 'historical, plain method', as opposed to Descartes's rigidly deductive system. In his *Essay Concerning Human Understanding*, published in 1690, he took the individual contents of the mind one by one and examined them to see what they were. The whole of knowledge, he argued, consisted of the collection and comparison of ideas. His was an empirical approach which stressed the role of experience rather than logical deduction in the acquisition of knowledge. It must not be thought, however, that the influence of *l'esprit géométrique* was entirely lacking in Locke's work. On the contrary, he took from Descartes his confidence in clear and distinct knowledge, and the *Essay* itself was written in prose of the utmost clarity which bears witness to the influence of scientific ideals of rigour and simplicity on Locke's style and vocabulary.

SCIENCE AND LITERATURE

The profound general influence of these mathematical ideals on literary style can be illustrated from developments in England in the seventeenth century – compare for example the work of the metaphysical poets at the beginning of the century with the much simpler and more lucid verse of Dryden at the end. But the influence was most obvious in a new attitude towards the writing of prose which had become common by 1700.

At the beginning of the seventeenth century Francis Bacon had stated that 'the ill and unfit choice of words wonderfully obstructs the understanding', and had urged the need for plain and precise prose, especially in scientific exposition. He himself did not always adhere to his own doctrines – his writings sometimes have an almost poetic strain – and early seventeenth-century prose, the prose of Burton, Browne and Milton, though it produced some magnificent passages which reflected the genius of their authors, was not notable for its clarity and simplicity. With the Restoration, however, came the foundation of the Royal Society, whose members took up Bacon's point about the need for a plain prose style. Thomas Sprat, the Society's first historian, wrote in 1667 that:

> There is one thing more about which the Society has been most solicitous, and that is the manner of their discourse: which, unless they had been only watchful to keep in due temper, the whole spirit and vigour of their design had been

soon eaten out by the luxury and redundance of speech. . . . And, in a few words, I dare say that of all the studies of men, nothing may be sooner attained than this vicious abundance of phrase, this trick of metaphors, this volubility of tongue, which makes so great a noise in the world. . . . It will suffice my present purpose to point out what has been done by the Royal Society towards the correcting of excesses in natural philosophy, to which it is of all others a most professed enemy. They have, therefore, been most vigorous in putting in execution the only remedy that can be found for this extravagance, and that has been a constant resolution to reject all amplification, digressions, and swellings of style. . . . They have exacted from all their members a close, naked, natural way of speaking, positive expressions, clear senses . . . , bringing all things as near the mathematical plainness as they can.

Many contemporary writers were convinced by such arguments, among them Joseph Glanvill, born in 1636 and educated at Exeter and Lincoln Colleges, Oxford. In 1661 he published his *Vanity of Dogmatizing*, an attack on scholastic philosophy and in favour of the experimental approach in scientific investigation. The work was designed to break down opposition to new ideas and open men's minds to new possibilities, and in 1665 Glanvill, who had been elected Fellow of the Royal Society during the previous year, republished it, with some minor revisions and a dedication to the Society, under the title *Scepsis Scientifica*. A third edition, this time completely revised and much shorter, appeared in 1676 and in this final version Glanvill showed the extent to which he had been converted to the Society's scientific ideals. His original version of 1661 was written in highly ornate prose, but fifteen years later these excesses of language had largely disappeared. In 1676, for example, the 'praeterlapsed ages' of the first edition became 'past ages' and 'preponderate much greater magnitudes' was altered to 'outweigh much heavier bodies'.

113 Frontispiece of Thomas Sprat's *History of the Royal Society* (1667). Lord Brouncker, the Society's first president, is seated to the left of the bust of Charles II, Francis Bacon to the right.

Chap. XXV.

(3.) *We cannot know any thing in Nature without knowing the first springs of Natural Motions ; and these we are ignorant of. (4.) Causes are so connected that we cannot know any without knowing all ; declared by Instances.*

But (3.) we cannot know any thing of Nature but by an *Analysis* of it to its *true initial causes* : and till we know the first springs of natural motions, we are still but Ignorants. These are the *Alphabet* of Science, and Nature cannot be *read* without them. Now who dares pretend to have seen the *prime motive causes*, or to have had a view of *Nature*, while she lay in her *simple Originals?* we know nothing but *effects*, and those but by our *Senses.* Nor can we judge of their *Causes*, but by proportion to palpable causalities, conceiving them like those within the sensible *Horizon.* Now t'is no doubt with the considerate, but that the *rudiments* of Nature are very unlike the grosser *appearances.* Thus in things obvious, there's but little resemblance between

tween the *Mucous sperm,* and the compleated *Animal.* The *Egge* is not like the *oviparous* production : nor the corrupted *muck* like the *creature* that creeps from it. There's but little similitude betwixt a *terreous humidity,* and *plantal* germinations ; nor do *vegetable* derivations ordinarily resemble their *simple seminalities.* So then, since there's so much dissimilitude between *Cause* and *Effect* in the more palpable *Phænomena* , we can expect no less between them, and their *invisible* efficients. Now had our Senses never presented us with those obvious *seminal* principles of apparent generations, we should never have suspected that a *plant* or *animal* could have proceeded from such unlikely *materials* : much less, can we conceive or determine the uncompounded *initials* of natural productions, in the total silence of our Senses. And though the Grand Secretary of Nature, the miraculous *Des-Cartes* have here infinitely out-done all the Philosophers went before him, in giving a particular and *Analytical* account of the *Universal Fabrick* : yet he intends his Principles but for *Hypotheses,* and never pretends that things are really or necessarily, as he hath supposed them : but that they may be admitted pertinently to solve the *Phænomena,* and are convenient supposals for the *use of life.* Nor can any further account be expected from humanity, but how things possibly *may have been made* consonantly to sensible nature : but infallibly to determine how *they truly were effected,* is proper to him only that saw them in the *Chaos,* and fashion'd them out

X2   of

114   Two pages from the 1665 edition of Joseph Glanvill's *Scepsis Scientifica*. Note the ornate prose style.

There can be no doubt, in fact, that the scientific writings of the seventeenth century exercised a very important and lasting influence on English prose style and vocabulary, encouraging the use of abstract terms and of metaphors drawn from physical science and helping to promote the writing of that relatively clear and simple prose which has remained an ideal from the late seventeenth century until our own time.

This ideal affected the spoken as well as the written word, a development which can be seen in the changing style of preaching. In the earlier years of the century sermons, like those of John Donne, reflected the elaborations which were characteristic of the general prose style of the time. But by the end of the century the tone was being set by John Tillotson, who became Archbishop of Canterbury in 1691. 'He was not only the best preacher of the age', wrote Gilbert Burnet, a distinguished contemporary observer, 'but seemed to have brought preaching to perfection. His sermons were so well heard and liked and so much read, that all the nation proposed him as a pattern and studied to copy after him.' Tillotson, to quote Paul Welsby, a modern authority, 'reformed the style of preaching and transformed its content. In place of involved elaborate discourses, teeming with metaphor and imagery, his sermons displayed ease of delivery, clarity and simplicity.'

Although English literary style was not substantially affected by the new scientific spirit until the later years of the seventeenth century, the content of literary works was influenced a good deal earlier by the new discoveries. It is true that Copernican ideas made only slow progress in Elizabethan England, even among the learned – Elizabethans' view of the universe, as represented by the cosmology of their dramatists and poets, was essentially Aristotelian and Ptolemaic – but there was a dramatic change in the early seventeenth century when Galileo's astronomical discoveries excited the imagination of the thinkers and writers of Europe. English poets responded almost at once. In 1611, only a year after the publication of Galileo's *Sidereus Nuncius*, John Donne produced his *First Anniversary*, a poem in which he set out the malaise produced by the spread of Copernican and Galilean ideas and the beginnings of the downfall of the old cosmology.

> *And new Philosophy calls all in doubt,*
> *The Element of fire is quite put out;*
> *The Sun is lost, and th'earth, and no man's wit*
> *Can well direct him where to look for it.*
> *And freely men confess that this world's spent,*
> *When in the Planets, and the Firmament*
> *They seek so many new . . .;*
> *'Tis all in pieces, all coherence gone;*
> *All just supply, and all Relation.*

Galileo's great discoveries in the heavens were made possible by the invention of the telescope, and references to the telescope

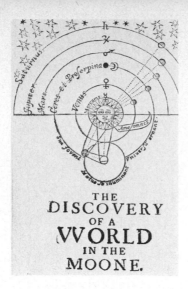

THE
DISCOVERY
OF A
WORLD
IN THE
MOONE.

115 Title-page of a book by John
Wilkins, one of the founders of the Royal
Society, who speculated on the possibility
that the moon was inhabited.

multiplied rapidly in the work of English poets, where it
became known by a variety of names: the 'perplexive glass', the
'optic magnifying glass', the 'trunk-spectacle' or 'trunk', the
'perspective', the 'glass'. Most frequently of all it was the 'optic
tube' or merely the 'tube'. Milton made several allusions to the
telescope in *Paradise Lost* and *Paradise Regained*. In the former
work there is a reference to the 'glass of Galileo', a comparison
of Satan's shield with the 'optic glass' of the 'Tuscan artist' and
the suggestion that the Garden of Eden was

> *a spot like which perhaps*
> *Astronomer in the sun's lucent orb*
> *Through his glazed optic tube yet never saw.*

In *Paradise Regained*, when Satan displays to Christ the kingdoms
of the world, Milton, speculating about the means which the
Devil employed to produce this remarkable vision, concludes,

> *By what strange parallax or optic skill*
> *Of vision, multiplied through air or glass*
> *Of telescope, were curious to inquire.*

In *Paradise Lost* Milton included descriptions of both the old
Ptolemaic and the new Copernican cosmology, but he made
no explicit choice between them. The most important effect of

162

116 The seventeenth-century gentleman was expected to be a person of versatile
accomplishment, interested in the sciences as well as in the arts, music and
literature. Detail from a painting showing the interior of a picture gallery. ▶

117 The poet John Milton (1608–74): he describes both th Ptolemaic and Copernican systems in *Paradise Lost*, and visited Galileo during a journe to Italy in 1638.

the new astronomy and more especially of the invention of the telescope upon Milton's mind was that it helped to give him a sense of vast distances in a virtually boundless universe: a sense of perspective and space which had no parallel in the work of earlier poets. There is a marvellous sense of cosmic perspective in the scene in *Paradise Lost* in which Satan

> *Looks down with wonder at the sudden view*
> *Of all this World at once. . . .*
> *Round he surveys (and well might, where he stood*
> *So high above the circling canopy*
> *Of night's extended shade) from eastern point*
> *Of Libra to the fleecy star that bears*
> *Andromeda far off Atlantic seas*
> *Beyond the horizon; then from pole to pole*
> *He views in breadth, – and, without longer pause,*
> *Down right into the World's first region throws*
> *His flight precipitant, and winds with ease*
> *Through the pure marble air his oblique way*
> *Amongst innumerable stars, that shone*
> *Stars distant, but nigh-hand seemed other worlds.*

Milton's fascination with space can be seen in virtually every book in *Paradise Lost*: for example, his description of chaos

> *Before their eyes in sudden view appear*
> *The secrets of the hoary Deep – a dark*
> *Illimitable ocean, without bound,*
> *Without dimension; where length, breadth, and highth,*
> *And time, and place are lost*

and the episode from the story of the creation in which Christ and his attendant angels survey the scene upon which God is about to impose order:

> *On heavenly ground they stood, and from the shore*
> *They viewed the vast immeasurable abyss,*
> *Outrageous as a sea, dark, wasteful, wild,*
> *Up from the bottom turned by furious winds*
> *And surging waves.*

Such preoccupation with the vastness of the universe is totally lacking in Shakespeare, whose world was still bounded by the sphere of the fixed stars, but Shakespeare, unlike Milton, had never looked through a telescope.

The new science is reflected in the content of prose as well as poetry, and Robert Burton's 'Digression of the Air' in his *Anatomy of Melancholy*, first published in 1621, is the liveliest and fullest survey in early seventeenth-century literature of the numerous conflicting astronomical ideas of the time: theories, for example, about the problem of change and decay in the heavens, about the movements of the heavens and the heavenly bodies, and about 'that main paradox of the earth's motion, now so much in question'. Burton found himself 'almost giddy with roving about', but this did not prevent him from greatly extending the passage on astronomy in the five revised editions of his book, the last of them coming out after his death in 1640.

The 'Digression of the Air' must have had considerable educative value for the small minority of Englishmen who read

118 (Overleaf) Jan Brueghel's *Sight*, painted in the early seventeenth century. The scientific instruments on the left bear witness to the growing effects of contemporary science on the imagination of artists.

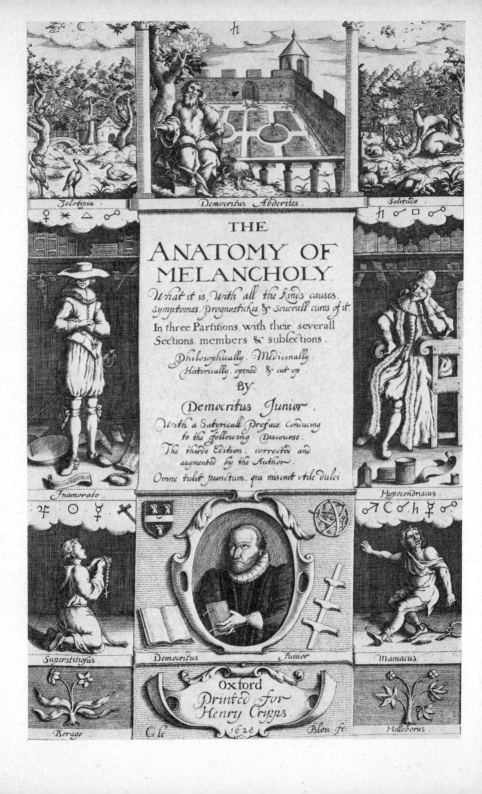

THE
ANATOMY OF
MELANCHOLY.
What it is, With all the kinds causes,
symptomes, Prognostickes, & seuerall cures of it.
In three Partitions, with their seuerall
Sections, members & subsections.
Philosophically, Medicinally,
Historically, opened & cut vp.
BY
Democritus Junior.
With a Satyricall Preface, Conducing
to the following Discourse.
The thirde Edition, corrected and
augmented by the Author.
Omne tulit punctum, qui miscuit vtile dulci.

Zelotypia.   Democritus Abderites.   Solitudo.

Inamorato.   Hypocondriacus.

Superstitiosus.   Democritus Junior.   Maniacus.

Borage.   Oxford
Printed for
Henry Cripps
1628.   Blon fe.   Helleborus.

20 Jonathan Swift, whose *Gulliver's Travels* ridiculed the ideas and experiments of scientists.

it, but it is worth noting that Burton himself refused to choose among the conflicting theories which he expounded. His whole approach was interrogative, even ironic, and that sceptical attitude towards the new science was shared by Jonathan Swift, one of the most distinguished English writers of the late seventeenth and early eighteenth centuries.

Swift was engaged on his most famous work, *Gulliver's Travels*, from at least 1720 until its appearance in 1726. He received a good deal of the inspiration for the book from the *Philosophical Transactions of the Royal Society*; a great amount of space in the *Philosophical Transactions* – especially in the volumes for the years 1700 to 1720 – was devoted to accounts of travel

◀ 119   The frontispiece from the third edition (1628) of Robert Burton's *Anatomy of Melancholy*.

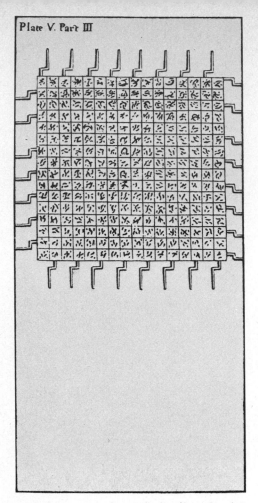

Plate V. Part III

121 An illustration, from *Gulliver's Travels*, of the word-making machine in the Grand Academy of Lagado. Using 'the most ignorant person, at a reasonable charge, and with a little bodily labour, may write books in philosophy, poetry, politics, law, mathematics, and theology, without the least assistance from genius or study.' Swift probably intended this as a satire on contemporary schemes for providing short cuts to knowledge by mechanical means.

and these may have helped to lead Swift to the general idea of *Gulliver's Travels*. That is conjecture, but it is certain that the *Philosophical Transactions*, together with the more complete works of the Fellows of the Royal Society, were the specific source for one of the notable episodes in Swift's book, the 'Voyage to Laputa'. In this story the flying island of Laputa hovered over the realm of Balnibarbi, which was ruled by the King of Laputa and had a capital city called Lagado, containing a 'Grand Academy'. Swift used his description of the Grand Academy of Lagado to mount a savage satirical attack on scientists in general and on the Royal Society in particular. Many of the experiments of the Academicians of Lagado which Swift described – and they included projects for extracting sunbeams from cucumbers, for converting ice into gunpowder, for building houses from the roofs downwards, and for preventing the growth of wool upon sheep – were based on the *Philosophical Transactions*. Swift, in fact, like so many laymen of the time, was convinced of the 'uselessness' of the scientific and mathematical learning which had developed so rapidly in the seventeenth century and had received widespread general publicity with the founding of the Royal Society. In this connection the section of the 'Voyage to Laputa' in which he discusses the mathematical obsessions of the Laputans is of considerable interest. The King of Laputa, observing the poor condition of Gulliver's clothes, ordered him to be measured for a new suit. The next morning a tailor appeared. Gulliver stated that

> This operator did his office after a different manner from those of his trade in Europe. He first took my altitude by a quadrant, and then with rule and compasses, described the dimensions and outlines of my whole body, all which he entered upon paper, and in six days brought my clothes very ill made, and quite out of shape, by happening to mistake a figure in the calculation. But my comfort was, that I observed such accidents very frequent, and little regarded.

In another passage, in which he described the houses of the Laputans, Gulliver, echoing Swift's own ideas, observed that these were

> very ill built, the walls bevel, without one right angle in any apartment, and this defect ariseth from the contempt they bear to practical geometry, which they despise as vulgar and mechanic, those instructions they give being too refined for the intellectuals of their workmen, which occasions perpetual mistakes. And although they are dexterous enough upon a

171

La Piramide A B C

Kheum, kheum, kheum;

La Piramide A B C D eſt donnée. Tout le corps contient 25. cubes, les lignes A D γ 160$\frac{52}{100}$ B D γ 160$\frac{33}{100}$ C D γ 162$\frac{1}{4}$.

Je demande les trois lignes de la baze.

O Soudari Bertrand *dit*.

He *Monſeigneur* notre maitre vous leurs auez tout dit!

Monſeigneur le Directeur *repond*.

He, mes chers enfans, je ne donne jamais des matieres mal aiſées a comprendre, je vous diray qu'au bout de notre rue il y a un tonnelier qui n'a jamais apris les matematiques, & n'eſt point penſionnaire de l'Academie Royale, & toutefois il ſçait faire le fons d'une tonne, auſſi-

auſſibienque celuy d'un cuuier, c'eſt pourquoy il ſera aiſé a ces Meſſieurs de l'Academie Royale de trouuer le fons d'une Piramide.

Autre Propoſition de Monſeigneur le Directeur.

La Piramide triangulaire A B C D eſt donnée, tout le corps contient 49. cubes, les lignes A D γ 445$\frac{51}{96}$ B D γ 448$\frac{5}{804}$ C D γ 454$\frac{47}{96}$.

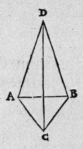

Je demande la hauteur & le point de la perpendiculaire,

B 3   *Moſci*

122   Two pages from *Le Reueil-matin*, a satire on the proceedings of the Académie Royale des Sciences, Paris, published in 1674.

piece of paper in the management of the rule, the pencil and the divider, yet, in the common actions and behaviour of life, I have not seen a more clumsy, awkward and unhandy people.

Finally, in the account of the foundation of the Grand Academy at Lagado, which Swift described through Gulliver, he revealed his fears of the disastrous effects which the wilder schemes of scientific enthusiasts might produce.

About forty years ago certain persons . . . fell into schemes of putting all arts, sciences, languages, and mechanics upon a new footing. To this end they procured a royal patent for erecting an Academy of Projectors in Lagado; and the humour prevailed so strongly among the people that there is not a town of any consequence in the kingdom without such an Academy. In these colleges, the professors contrive new rules and methods of agriculture and building, and new instruments and tools for all trades and manufactures, whereby, as they undertake one man shall do the work of ten, a palace may be built in a week, of materials so durable as to last for ever, without repairing. All the fruits of the earth shall come to maturity at whatever season we think fit to choose, and increase a hundred fold more than they do at present; with innumerable other happy proposals. The only inconvenience is, that none of these projects are yet brought to perfection, and in the meantime the whole country lies miserably waste, the houses in ruins, and the people without food or clothes.

These denunciations of the exaggerated expectations arising from the new science were echoed in other quarters, by Joseph Addison, for example, who used his famous newspaper, the *Spectator*, which appeared in 1711–12 and again in 1714, to poke fun at absent-minded and impractical scientists who were so preoccupied with their subject that they forgot the realities of the world about them.

THE SPREAD OF SCIENCE

The attacks of Swift, Addison and others, though they gave scientists a bad press, did, however, show that scientific ideas and discoveries were, by the end of the century, of widespread interest to the educated classes in England, who read not only such satirical works but also the growing flood of books in which the scientists themselves expounded and sometimes even attempted to popularize their ideas. Indeed, this developing interest in scientific books was a European phenomenon, and between 1600 and 1700 the literate sections of European society, at least in the western half of the Continent, absorbed from such works the basic ideas of the scientific revolution. This in turn helped to produce the tremendous secularization of thought which characterized the last decades of the seventeenth century and the first decades of the eighteenth – the change in intellectual assumptions of a large part of the educated classes which the distinguished French historian Paul Hazard aptly christened *la crise de la conscience européenne.*

During the seventeenth century an ever-increasing spate of books on travel poured from the printing presses. These accounts covered every continent of the globe and by the end of the century they were beginning to exercise a profound influence on the minds of Europeans, who became familiar with the ideas and customs of peoples, such as the Chinese, who knew little of the classical civilizations of the ancient world or of Christianity. This raised the possibility that European customs were no better than those of other parts of the earth's surface, and even, and much more dangerously, the thought that the Christian belief in God was only one of a number of alternative acceptable conceptions of the Deity. Here we can see, in unmistakable form, one of the roots of eighteenth-century deism.

It was in the late seventeenth century, too, that the great French writer, Pierre Bayle, stressed more clearly and cogently than anyone had ever done before the idea that there was no necessary connection between morality and the Christian religion, publicizing his theories in such works as his *Pensées sur la Comète*, which he produced in 1682, and in his famous *Dictionnaire historique et critique*, published in the closing years of the century. In this Bayle pointed out that many orthodox Christians lived very unedifying lives, giving themselves over to such unattractive practices as murder, rape and fraud. On the other hand there were plenty of freethinkers – Spinoza is an example – of the loftiest moral character. The conclusion was clear.

174   Morals and religion [Bayle affirmed], far from being in-

123 Travel books made literate Europeans increasingly aware of the geography and culture of newly discovered regions of the world. This illustration, taken from a French book published in 1578, shows a family of Greenlanders captured by the English explorer Martin Frobisher (1535–94).

separable, are completely independent of each other. A man can be moral without being religious. An atheist who lives a virtuous life is not a creature of wonder, something outside the natural order, a freak. There is nothing more extraordinary about an atheist living a virtuous life, than there is about a Christian leading a wicked one.

Such a statement must have horrified both the Catholic and Protestant churches. It seems clear, however, that at the end of the seventeenth and the beginning of the eighteenth centuries there was a decline in the strength of institutional Christianity which helped the spread of Bayle's scepticism, but was only partly caused by it and by the increasing influence of the ideas of the scientific revolution. In England the failure to comprehend nonconformists within the restored Anglican Church of 1660 and the passage of the Toleration Act of 1689 meant the end of the ideal of a Christian English commonwealth in which all the citizens were members of the Church as well as the State. This weakened the position of the Church of England as an institution at a time when it was still trying to recover its moral stature after the startling events of the 'Puritan revolution' of 1640–60. In France, too, the institutional authority of the Gallican Church was undermined by the conflicts between the Papacy and Louis XIV on the one hand, and between the 'orthodox' and Jansenist factions within the Church on the other, while with the beginnings of the Regency in 1715 there came unmistakable signs of contempt for organized religion among the upper classes.

Scientific ideas at the turn of the seventeenth century, with their stress on the need for rational explanations, led to a rejection of belief in concepts, such as miracles, witchcraft and astrology, which could not be explained in scientific terms, and to the weakening of the belief of many educated men in the traditional doctrines of Christianity. The new scientific ideas were spread by the publications of learned societies and by the increasing number of scientific books which were so avidly read by the educated members of western European society during the seventeenth century.

One man, however, deserves special mention as the greatest popularizer of the scientific discoveries of the time. That is Bernard de Fontenelle, who was born in 1657 and died in 1757, only one month before his hundredth birthday. Through his writings and his work between 1691 and 1741 as secretary of the Académie Royale des Sciences, he made, in clear and simple terms, many of the scientific advances of the seventeenth century known to the reading public, and especially to the

124   Frontispiece of
The Discovery of Witches (1647) by
Matthew Hopkins, showing two witches
and their familiars. Hopkins, a
notorious detector of witches,
was eventually totally discredited.

increasingly numerous and influential French bourgeoisie. Above all, because of his own opinions, Fontenelle was the most important single link between the scientific movement of the seventeenth century and the *philosophes* in the eighteenth. Many of the leading seventeenth-century scientists were pious Christians, but Fontenelle was a sceptic, and, in transmitting the ideas of the great scientists for general consumption, he gave them an anti-religious slant; this helped to strengthen the already existing impression that the Church, especially in France, was the enemy of scientific advance. It was this attitude which Fontenelle transmitted to his intellectual heirs, such as Voltaire, whose views both favoured the spread of scientific knowledge and challenged the truth of revealed religion.

The scientific advances of the seventeenth century thus had a far-reaching effect upon the minds of literate men in western Europe, imbuing many of them by 1700 with a rational and sceptical outlook on life which rejected mysterious explanations of natural phenomena and questioned the axioms of revealed religion. An intellectual gulf had developed not only between western and eastern Europe – Russia was quite unaffected in the seventeenth century by the scientific revolution – but also between the educated élites of the west, especially in France and England, and their uneducated fellow countrymen. There is no evidence that, by 1700, the new scientific ideas had had any effect on the minds of French and English peasants, who probably still regarded the universe as earth-centred and certainly continued to believe in the reality of witchcraft and astrology and in the truth of orthodox Christian tenets. On the other hand, by the end of the seventeenth century educated Frenchmen and Englishmen usually held Copernican views about the structure of the universe, perhaps no longer believed in witchcraft or astrology, and might question or reject some of the basic dogmas of Christianity. This breach between the assumptions of the intelligentsia and the uneducated was far more profound than any changes in ideas about God and man which the Protestant Reformation of the previous century had brought about. The Reformation had not affected such Christian fundamentals as the belief that Christ was a Man–God who had died on the Cross to save all mankind, nor had it produced an obvious division between the conceptions which the educated and the ignorant held about the universe and man's place in it. It took the intellectual developments of the seventeenth century – with the scientific revolution at their centre – to widen the cultural gap between the intelligentsia and peasantry of western Europe to an extent which left them living, in 1700, in different mental worlds. It is, in fact, among the educated élite of the later part of the century that we first discern mental attitudes, such as scepticism about certainties founded on faith rather than on reason, which are recognizably modern. In the field of general intellectual assumptions it is this period at the end of the seventeenth and the beginning of the eighteenth centuries, roughly the years between 1680 and 1720, and not the era of the Reformation, that really marks the transition from the 'middle ages' to 'modern times'.

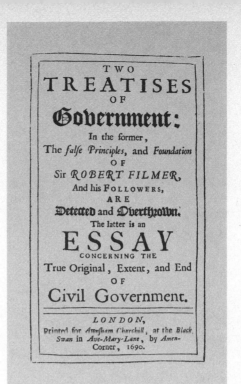

TWO
TREATISES
OF
Government:
In the former,
The *false Principles*, and *Foundation*
OF
Sir *ROBERT FILMER*,
And his FOLLOWERS,
ARE
Detected and Overthrown.
The latter is an
ESSAY
CONCERNING THE
True Original, Extent, and End
OF
Civil Government.

LONDON,
Printed for *Awnsham Churchill*, at the *Black*
Swan in *Ave-Mary-Lane*, by *Amen-*
Corner, 1690.

125   Title-page of John Locke's
Two Treatises of Government, dated 1690
but actually published the year before,
when William III succeeded
to the throne.

SCIENCE AND GOVERNMENT

The rational ideas characteristic of the scientific revolution,
which affected so many areas of human thought, almost in-
evitably had an impact on problems of government and
administration. John Locke, who was a political theorist of the
first importance as well as a general philosopher, produced in
1689 his *Two Treatises of Government*, in the second of which
he rejected ideas of divine right and envisaged the basis of civil
government in a 'reasonable' social contract in which the
generality of men gave up their equality and 'natural' executive
power in order to preserve their liberty and property. Rulers,
therefore, were under an obligation to conduct government in
the interests of their subjects, and, if they failed to do so, then
the people were entitled to rebel and expel them from authority.
Locke's ideas had a powerful influence in eighteenth-century

179

Europe. He was the ancestor of much of the rationalism which characterized the political thought of the *philosophes* and which rejected the inherited and seemingly unreasonable powers and privileges of kings, clergy and nobility in favour of an empirical approach to the problems of government. The ideas of science, when transmitted to the political sphere, were thus inimical to the whole governmental structure – in its widest sense – of the *Ancien Régime* and helped to prepare the way for the upheavals of the French revolutionary period.

Scientific developments also began, in this period, to have an effect on the administrative machinery through which governmental decisions were carried out. The second half of the seventeenth century saw the rise of what contemporaries called political arithmetic and we would call statistics, and it was quickly realized that statistical methods – themselves a reflection of the age's rapidly developing interest in all branches of mathematics – would, if they could be applied to problems of administration, vastly increase a government's knowledge of the men and natural resources at its disposal and consequently improve its efficiency and its potential political power *vis-à-vis* its neighbours.

Attempts to use statistics for governmental purposes had, of course, a very long ancestry; indeed, the Roman administration showed a sophisticated interest in such matters. But there was a notable increase of interest in this quantitative approach, particularly as regards trade, in much of western Europe in the later sixteenth century. This development can be explained by the growing knowledge of at least elementary arithmetic among the governing classes as well as by the increasing complexity of the social and economic structure of the European states, which made precise knowledge about their economic and financial conditions more necessary as well as more desirable for governments.

It was only in the second half of the seventeenth century, however, that the quantitative approach to administrative problems began to develop in a fairly sophisticated way with

126 An eighteenth-century orrery, a mechanical model used
for showing the movements of the planets round the sun.
Detail from a painting by Joseph Wright of Derby. ▶

the work of John Graunt, William Petty and Gregory King and their followers in England, and the labours of Colbert, Vauban and others in France. During the later years of the century, and particularly after 1689, the governments of England and France were faced with economic, military and financial problems of hitherto unparalleled magnitude; these were caused, above all, by the great wars of the period and were characterized by a dramatic expansion in the numbers of state administrators who were required to deal with such problems. These developments, which reproduced on a much larger scale the situation of the later sixteenth century, made the necessity for accurate statistical information about governments' resources abundantly clear. This occurred at the very time when the continuing spread of mathematical knowledge and the development of techniques of statistical analysis made it possible to begin to satisfy governments' needs.

In France, in the middle of the seventeenth century, there already existed a considerable interest in statistical surveys, but this only really came to fruition in 1664 when Colbert, Louis XIV's minister, undertook an inquest which covered almost all the French provinces and which was carried out by the *intendants*, those invaluable local officials appointed by the central government, who became a permanent part of the French administrative machinery at this time. Colbert asked for good maps of the provinces, with the administrative divisions clearly marked; for details of the personnel and resources of the Church; for information about the nobility and the machinery of justice; for statistics about finance and the crown lands; for details about the natural resources of the provinces and about the activities of the inhabitants; and for information about the state of commerce and industry. These were only the principal demands which he made upon his officials and, although the inquest of 1664 was never fully completed, the work was taken up later in the reign, especially in the 1680s, by the military engineer Sebastien Vauban, who, in his travels through France, left questionnaires containing requests for

127   Bust by Coysevox of Jean–Baptiste Colbert, Louis XIV's finance minister, an administrative reformer and founder of the French Academy of Sciences. ▶

à l'Egard
de l'Eglise.

Le nom et le nombre de ses
éueschez, les villes, Bourgs, et
Bourgades et parroisses qui sont
soumises à la Jurisdiction eclesi-
astique.

Leurs seigneuries temporelles
et les villes et parroisses dont
elles sont composées, particuliere-
ment si l'Euesque est seigneur
temporel de la ville cathedralle.

Le nom, âge, et l'état de la
disposition de l'Euesque.

S'il en dupayee ou non.

S'il y fait residence ordinaire.

De quelle sorte il s'aquitte de ses
visittes.

Quel credit il a dans son pays,
et quel effet il pourroit faire dans
les temps difficiles.

En quelle reputation il est
parmy les peuples.

S'il conferé les benefices de son
chapitre.

S'il est en procez auec son
chapitre, son revenu.

Et le nom et valeur des benefices
qu'il confere.

Outre ce qui concerne les eueschez,
et tout ce qui en depend, il est
necessaire de sçauoir le nom et le

128 Instructions sent out for the French census of 1697, asking for precise information about ecclesiastical personnel, possessions and jurisdiction.

statistical information in the hands of the *intendants*. As a result,
he received many detailed reports, and Fontenelle, who had
the greatest regard for him, said, with some exaggeration, that
the development of statistical analysis was virtually due to
Vauban alone.

The last great inquiry of Louis XIV's reign, and unquestionably the best known, was set on foot in 1697 in order to provide
the Duc de Bourgogne, Louis XIV's grandson, with an accurate
account of the state of the kingdom. The instructions sent to
the *intendants* made it very clear that they were to provide as
many statistics as they possibly could. Most of the replies, not
all of which were drawn up by the *intendants* themselves, were
finished by the early months of 1698, but the young Duke, who
was given such a splendid introduction to the condition of the
kingdom which he was one day expected to rule, was never

able to put any of the information to effective use. He died in 1712, three years before Louis XIV.

In England also, the later seventeenth century saw important advances in the application of statistics to administrative and social problems. Sir William Petty, one of the early Fellows of the Royal Society, stood at the centre of these developments. Petty was something of a polymath – doctor, surveyor, geographer, political economist, statistician, civil servant, and an able businessman to boot! He was firmly convinced that the quantitative approach was the key to an understanding of social and administrative problems. Statistics were indispensable 'in order to get good, certain and easy government'. Petty was, in the words of his cousin, Sir Robert Southwell, 'ravished with the harmony and charms of ratiocination' and he co-operated with John Graunt in the production in 1662 of *Observations upon the Bills of Mortality*, a work which included information about the numbers of deaths in London caused by various diseases, and which was the first genuinely statistical work ever written in England. Five years later Petty wrote his *Verbum Sapienti*, a work designed to influence tax policy by showing that it would be advantageous to tax people more heavily and property less heavily. If his arguments were to

IN Parishes of about an Hundred Families, and wherein the Registry of the Births, Burials, and Marriages hath been well kept, Enquire,

1. The Number of the Inhabitants, Male and Female.
2. Married and Unmarried, and their Trades.
3. Widdows and Widdowers.
4. The Age of each Person, Man, Woman, and Child.
5. The Number of Families and Hearths.

As in the following Scheme, *Viz.*

Hearths.	Males.	Females.
3.	*John Smith*, Taylor, 45. 17. 15. 1.	His Wife 40. 16. 14. 4.
2.	*Richard Sims*, Carpenter, 52. 30. 22. 11.	His Wife 46. 24. 12. 2.
4.	*Robert Hughs*, Shoemaker, Widdower, 50. 16. 14. 2.	18. 6. 1.

And put the Births, Burials, and Marriages into the following Scheme, for the 7 last Years.

	Born.		Buryed.		Married.
	Males	Females.	Males	Females.	
Anno 1676.					
77.					
78.					
79.					
80.					
81.					
82.					

Describe the Soyl and Scituation of the Parish, and the Reputed Number of Acres which it containeth.

129 An anonymous English census form of the late seventeenth century.

carry conviction, it was essential that they should be based on statistics, and the result was the first serious attempt to calculate the national income. The inhabitants of England, he believed, spent about £40 million per annum. Some £25 million of this derived from the labour of individuals, whereas only £15 million was income from property – figures which supported his idea about the benefit to the State of relatively higher personal taxation.

Petty's most distinguished immediate successor was Gregory King, who completed his *Natural and Political Observations* in 1696. This work, the most important statistical survey of the time, was not published in full until the beginning of the

130 An illustration (1684) of a fifth-rate ship: the classifying of ships, which was introduced by Samuel Pepys (1633–1703), one of the most important English

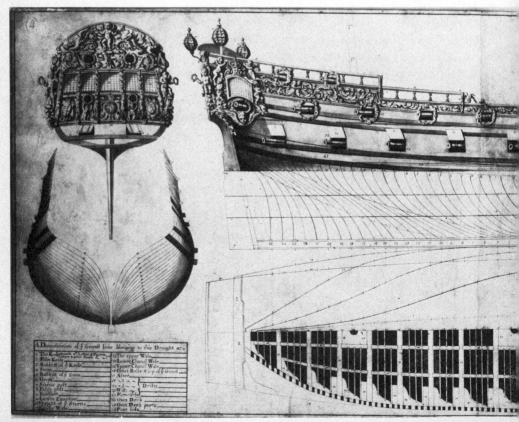

nineteenth century, but King's findings were quoted by another author as early as 1698. King wrote during the Anglo-French war of 1689–97 and he was well aware of the need for accurate information about the condition of the country at such a time.

If to be well apprized of the true state and condition of a nation, especially in the two main articles of its people and wealth, be a piece of political knowledge of all others and at all times the most useful and necessary, then surely at a time when a long and very expensive war . . . seems to be at its crisis, such a knowledge of our own nation must be of the highest concern.

administrators of the century, is an example of the growing desire to adopt a scientific approach to administrative, military and naval problems.

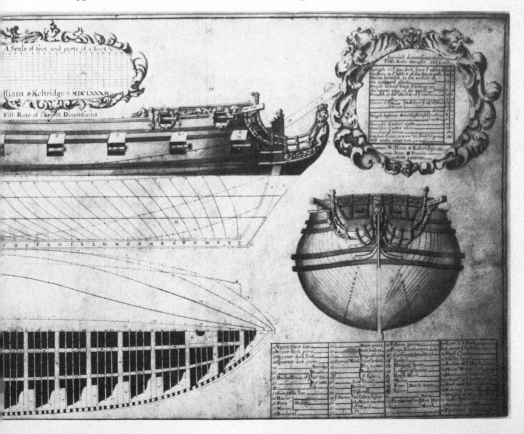

He then proceeded to divide the entire population of the country into its social groups, from the nobility down to the poor and vagrants, crediting each section of the community with the proportion of the national income which it represented. His work was based largely on a careful sampling of tax records and the accuracy of many of his conclusions has been highly praised by modern authorities.

The year in which King completed his remarkable work was, in fact, something of a landmark in the history of English statistical analysis, because it also saw the foundation of what has been described as 'the first special statistical department successfully created by any western European state'; the office of Inspector-General of Imports and Exports. The establishment of the new office was doubtless partly due to the current popularity of political arithmetic, but its immediate occasion was the financial crisis of 1695–96, one of England's worst moments in the war against France. This crisis also led to the foundation, again in 1696, of the Board of Trade, an institution which likewise reflected the prevalent belief in the value of statistics. John Locke was one of its members.

Five years after these events, a distinguished writer, John Arbuthnot, stated the case for statistics in his *Essay on the Usefulness of Mathematical Learning*.

> Arithmetic is not only the great instrument of private commerce, but by it are (or ought to be) kept the public accounts of a nation; I mean those that regard the whole state of a commonwealth, as to the number, fructification of its people, increase of stock, improvement of lands and manufactures, balance of trade, public revenues, coinage, military power by sea and land etc. Those that would judge or reason truly about the state of any nation must go that way to work, subjecting all the forementioned particulars to calculation. This is the true political knowledge. In this respect the affairs of a commonwealth differ from those of a private family only in the greatness and multitude of particulars that make up

the accounts. . . . What Sir William Petty and several others of our countrymen have wrote in political arithmetic does abundantly show the pleasure and usefulness of such speculations. It is true, for want of good information, their calculations sometimes proceed upon erroneous suppositions; but that is not the fault of the art. But what is it the government could not perform in this way, who have the command of all the public records?

This passage conjures up the rational administrative methods of the twentieth century and indicates that the later seventeenth century saw the tentative beginnings in western Europe of an administrative revolution, founded on the techniques of statistical analysis. After setbacks in the eighteenth century, these techniques developed over the years both in sophistication and in geographical extent until today there is no country in the world with any pretence to civilization that does not endeavour to apply statistics to administration. Thus, the application of rational, scientific techniques to administrative problems marked the start of a decisive breach with 'medieval' administrative methods and the recognizable beginnings of 'modern' ideas. In administration, just as in general intellectual assumptions, the last decades of the seventeenth century marked the start of a new historical era.

SCIENCE AND THE ECONOMY

The scientific advances of the seventeenth century also affected the economy and society of the day, though these effects were a great deal less striking than Francis Bacon would have liked to see. Bacon had assumed that advances in scientific knowledge would lead to increased power over nature and thus to an increasing ability to better man's lot on earth. Some of the scientific discoveries of the period did have practical social and economic effects. By about 1700, however, these had produced improvements only in the circumstances of the upper echelons of society. Bacon's optimism, as applied to the general run of

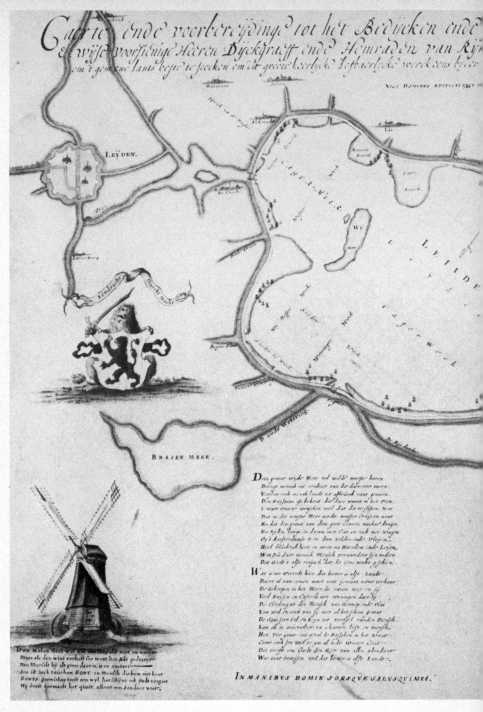

131   Plan for draining the Haarlemmermeer (1640): during the previous century, large-scale drainage schemes had been started in the Netherlands.

ken vande Haerlemer-Meer omte vertoonen aende
die selue met goeden raet en daet de hant daer mee willen aenhouwen
om te bedijcken ende met Godts hulpe tot goet lant te brengen, ende profijt deen. Anno 1640.

ANT QVI AEDIFICAT EAM. PSLM LXVII.

HAERLEM.

HAERLLMER-MEER.

SPIERING-MEER.

Amsterdammer Weech

D'OUDE-MEER.

Schildts toe weck

Luthe-Meer.

Sloten.

Rijnlandtsche Mate 12 voeten maken een Roede
ende 600 Roeden maken een Rijnlandtsche Morgen
ende jder parck in dese Caert maeckt 10 Morgen
en het quadraet van dien is 115 Roeden seggen.

Gedaen door Jan & Pieter Gael Water, Ingenieur ende
Water-Makers vande Rijp in Noordt-Hollandt
...

A Dieu Seul Honneur & Gloire.

Nieuwer Meer.

Het luchtse water   vant Haerlemes

AMSTERDAM.

De HAERLEMER MEER groot
bevinden omtrent 24000. Morgen
ende inden ompanck omtrent
...

L. AB OMNI PARTE BEATVM.

the population in western Europe, did not really bear fruit at this time.

A basic need, if scientific discoveries are to have a significant practical effect on the conditions of life of the average man, is a close, prolonged and large-scale alliance between science and technology. This did not exist in the sixteenth and seventeenth centuries and was not to exist before the later eighteenth century at the earliest – such attempts as were made in the seventeenth century to create links between science and technology, notably those made by the Royal Society and the Académie Royale des Sciences in the 1660s, were short-lived and on a comparatively small scale.

During the years between 1500 and 1700 better agricultural methods, themselves the product to some extent of more scientific procedures for surveying and draining estates, and a growth in trade which was certainly linked to both the geographical discoveries and scientific advances of the period, produced substantial benefits for the pockets and stomachs of landlords and merchants. At the same time, however, the lot of the peasants, even in western Europe, improved very slowly, if at all. During these centuries, therefore, such scientific developments as could be put to practical use were exploited by those sections – landowners and merchants – of the west European community which possessed political and economic power to increase their own wealth; but comparatively few of the benefits percolated down to the 'lower orders' of society.

It is worth noting, too, that by the later seventeenth century the countries which had the most prosperous economies, England, France and the United Provinces, were also the most advanced scientifically. During the two centuries in fact the economic and political centre of gravity of Europe moved from the Mediterranean to the Atlantic. This was one of the decisive developments in world history, and here, too, the scientific revolution clearly played its part.

Some *specific* results of scientific advances on a country's economy can be found in the history of the United Provinces

132   A sixteenth-century theodolite, the basic instrument used in land surveying.

during their 'golden age' in the seventeenth century. In the words of Charles Wilson, the Dutch economy owed an 'uncommon debt' to scientific developments. This was because its basis was seaborne trade, and scientific advances, especially in the fields of instrument-making and cartography, could make long voyages safer, quicker and hence more attractive. In the seventeenth-century Dutch Republic, indeed, there was a complicated relationship between scientific discovery and economic advance, each stimulating the other.

Dutch scientists and craftsmen were far ahead of all foreign rivals in the techniques of lens grinding and polishing, and the remarkable range of telescopes, binoculars and scientific instruments which they were able to produce was of vital help to Dutch navigators, who depended on at least an elementary knowledge of astronomy for success in their far-flung voyages. Accurate maps were another vital aid to navigation, and in this field, too, the Dutch excelled; almost all the best maps published in the sixteenth and early seventeenth centuries were

133 Dutch engraving satirizing the growing habit of wearing spectacles: the increasing use of glasses owed much to the skill of the Dutch lens-grinders and polishers.

produced by a school of cartographers which grew up in the Netherlands in the sixteenth century and is associated with such famous map-makers as Mercator, Ortelius, Wagenaer and Blaeu.

In discussing the more narrowly economic influence of the scientific revolution it is natural to concentrate on trade and navigation. It is perhaps right, when considering its wider social implications, to focus attention on medicine, a field where the practical application of improvements could do so much to alleviate individual human misery, and, if conditions were favourable, to improve the quality of life in the community as a whole. Unfortunately, however, many of the most important advances in knowledge, in particular the work of Vesalius and Harvey, had little immediate effect on medical practice. In this respect the work of Paracelsus and his followers, which introduced new chemical drugs and treated disease as a manifestation of disorders in individual organs rather than in the body as a whole, was probably a good deal more effective. All in all, because of the ideas of the Paracelsians and because of some general improvements in the medical education of sixteenth- and seventeenth-century doctors, individual patients who were subjected to medical ministrations had, perhaps, a marginally better chance than before of survival. The improvement was, however, slight at best, and anyone who reads accounts of the savage purgings and bleedings which were commonplace at the court of Louis XIV must marvel at the toughness of that monarch's constitution, which enabled him to survive to the age of seventy-six. His family were not all so lucky, and it is clear that the attentions of the royal doctors only served to contribute to the death from measles in 1712 of the elder of his

134   The agony of surgery without an anaesthetic is vividly conveyed in this early seventeenth-century painting. In the background, one patient is being bled, while the other has his urine examined.

135   (Overleaf) The siege of Namur (1695) by English and Dutch forces under William III. The increasing scale of warfare, which was an outstanding feature of the seventeenth century, eventually produced a much more scientific approach to the art of fortification and siege warfare.

195

great-grandsons. That child's younger brother, the future Louis XV, probably survived only because of the good sense of his nurse, who hid him away from the doctors' tender mercies.

In these centuries, too, social and economic developments – such as the vastly increasing size of towns, a dramatic growth in the scale of warfare both on land and at sea, and the introduction into Europe of new diseases such as scurvy and syphilis – greatly increased the incidence and risk of disease among the population. In the circumstances doctors had, as it were, to make substantial advances even to stand still, and it is clear that, despite the marginal improvements which were made in medical practice, they were unable to do so.

The scientific developments of the period did not, therefore, produce a significant betterment in the quality of life enjoyed by the generality of the population in early modern western Europe. The medical treatment which men received when ill, if they received any at all, remained primitive and frequently painful, and indeed the average expectation of life was probably shorter in the seventeenth century than in the previous one. The economic condition of the lower classes, the vast majority of them dependent on agricultural work, varied widely from year to year and from area to area and was determined largely by the vagaries of the annual harvest, but over the period as a whole it probably, on average, improved little if at all. The more prosperous sections of the community, notably the landlords and merchants, did, in contrast, increase their wealth substantially owing to a number of developments, among which scientific advances had their place. Thus, the scientific revolution, which helped to produce such a dramatic gap between the intellectual assumptions of the educated classes of western Europe and the bulk of the population, also helped to produce a growing gap between the economic élites of the area and ordinary men. In the widest sense it was a disruptive force in the society of the times.

C. S. Lewis, *The Discarded Image* (Cambridge 1964), is a brilliant discussion of the medieval model of the universe which was overthrown by the scientific revolution. Many general works contain good accounts of early modern scientific developments. Among the best are: W. P. D. Wightman, *The Growth of Scientific Ideas* (Edinburgh and London 1950); H. Dingle, *The Scientific Adventure* (London 1952); C. Singer, *A Short History of Scientific Ideas to 1900* (Oxford 1959); C. C. Gillispie, *The Edge of Objectivity* (Princeton 1960); E. J. Dijksterhuis, *The Mechanization of the World Picture* (Oxford 1961); A. C. Crombie, *Augustine to Galileo* (2nd edition, 2 vols. London 1961); J. H. Randall, *The Career of Philosophy from the Middle Ages to the Enlightenment* (New York 1962); A. R. Hall, *The Scientific Revolution* (2nd edition, London 1962); M. Boas, *The Scientific Renaissance 1450–1630* (London 1962); A. R. Hall, *From Galileo to Newton 1630–1720* (London 1963). J. D. Bernal, *Science in History* (3rd edition, London 1965), is a remarkable survey, written from a Marxist point of view, of the development of science from the earliest times to the twentieth century. R. Briggs, *The Scientific Revolution of the Seventeenth Century* (London 1969), is an interesting brief sketch, while G. N. Clark's masterly analysis, *The Seventeenth Century* (2nd edition, Oxford 1947), contains an excellent account of seventeenth-century science and its effects on contemporary life and thought. A. Koestler, *The Sleepwalkers* (London 1959), is an idiosyncratic but stimulating discussion of the work of Copernicus, Tycho Brahe, Kepler and Galileo. H. Kearney, *Science and Change 1500–1700* (London 1971), is an interesting general survey, and H. Butterfield, *The Origins of Modern Science* (new edition, London 1957), is a remarkable *tour de force*, which places sixteenth- and seventeenth-century scientific developments in their political and cultural setting. S. Toulmin and G. J. Goodfield discuss men's cosmological and astronomical theories from the earliest times to the days of Newton in *The Fabric of the Heavens* (Harmondsworth 1963), while C. Singer, *A Short History of Anatomy from the Greeks to Harvey* (London 1957), is a good introduction to medical ideas. Useful accounts of chemical developments can be found in F. J. Moore, *A History of Chemistry* (3rd edition, New York and London 1939); J. R. Partington, *A Short History of Chemistry* (2nd edition, London 1948); and H. M. Leicester, *The Historical Background of Chemistry* (New York and London 1956). A. O. Lovejoy, *The Great Chain of Being* (Cambridge, Mass. 1936), considers an intellectual conception which played a most significant part in European thought from the time of the Greeks until the eighteenth century. D. L. Hurd and J. J. Kipling (eds.), *The Origins and Growth of Physical Science* (2 vols. Harmondsworth 1964), and A. Rook (ed.), *The Origins and Growth of Biology* (Harmondsworth 1964), contain excerpts from some of the great seminal works of the scientific revolution.

H. F. Kearney (ed.), *Origins of the Scientific Revolution* (Problems and Perspectives in History, London 1964), is a useful collection of essays with an introduction by the editor, while G. N. Clark, *Science and Social Welfare in the Age of Newton* (2nd edition, Oxford 1949), contains, besides much else, a brief analysis of the origins of the scientific revolution. There has been a great deal of discussion about the general relationship between the development of capitalism and Protestantism on the one hand and the growth of early modern science on the other. Among those who see significant connections are C. Hill, *Intellectual Origins of the English Revolution* (Oxford 1965), 'Puritanism, Capitalism and the Scientific Revolution' in *Past and Present* no. 29, December 1964, 'Science, Religion and Society in the Sixteenth and Seventeenth Centuries' in *Past and Present* no. 32, December 1965; and S. F. Mason, 'Science and Religion in Seventeenth-century England' in *Past and Present* no. 3, February 1953. Among those who deny them are H. F.

Kearney, 'Puritanism, Capitalism and the Scientific Revolution' in *Past and Present* no. 28, July 1964; and T. K. Rabb, 'Religion and the Rise of Modern Science' in *Past and Present* no. 31, July 1965, and 'Science, Religion and Society in the Sixteenth and Seventeenth Centuries' in *Past and Present* no. 33, April 1966. B. J. Shapiro, 'Latitudinarianism and Science in 17th-Century England' in *Past and Present* no. 40, July 1968, stresses the role of religious moderates in the scientific revolution in England. J. H. Parry, *The Age of Reconnaissance* (London 1963), and J. R. Hale, *Renaissance Exploration* (London 1968), provide interesting accounts of the voyages of discovery which certainly gave *specific* stimuli to scientific advances. A. C. Crombie, *Robert Grosseteste and the Origins of Experimental Science 1100–1700* (Oxford 1953), and M. Clagett, *The Science of Mechanics in the Middle Ages* (Madison, Wisconsin 1959), stress the debt of the scientific revolution to medieval developments, as does E. A. Burtt in *The Metaphysical Foundations of Modern Physical Science* (London 1925), which emphasizes the contribution made by the Platonic revival of the later Middle Ages.

There are good discussions of sixteenth- and seventeenth-century mathematical advances in E. T. Bell, *The Development of Mathematics* (New York and London 1940); H. Eves, *An Introduction to the History of Mathematics* (revised edition, New York 1964); and D. J. Struik (ed.), *A Source Book in Mathematics* (Cambridge, Mass. 1969). F. E. Sutcliffe has translated *Descartes' Discourse on Method and Other Writings* (London 1968). A. Wolf, *A History of Science, Technology and Philosophy in the Sixteenth and Seventeenth Centuries* (2nd edition, London 1950), has a useful brief account of the development of scientific instruments, while M. Bronfenbrenner, *The Role of Scientific Societies in the Seventeenth Century* (Chicago 1928), considers the scientific journals as well as the societies of the period. T. S. Kuhn, *The Copernican Revolution* (Cambridge, Mass. 1957), is excellent on the significance of Copernicus's work, and J. L. E. Dreyer's *Tycho Brahe* (Edinburgh 1890) is still valuable. M. Caspar has written a useful account of *Johannes Kepler* (translated and edited by C. Doris Hellman, London and New York 1959), but the best account of Kepler is probably to be found in Koestler's *Sleepwalkers*. L. Geymonat, *Galileo Galilei* (translated by S. Drake, New York 1965) is a recent full-scale biography, and G. de Santillana, *The Crime of Galileo* (Chicago 1955), is the best book on Galileo and the Church. E. N. da C. Andrade, *Isaac Newton* (London 1950), is a good brief account of Newton's life and work, while the Royal Society's volume *Newton Tercentenary Celebrations* (Cambridge 1947) contains a brilliant essay by Lord Keynes on 'Newton the man'. A. Koyré, *Newtonian Studies* (London 1965), the work of a great scholar, contains some stimulating essays, while F. Manuel's recent book, *A Portrait of Isaac Newton* (Cambridge, Mass. 1968), is a massive Freudian study. F. R. Johnson, *Astronomical Thought in Renaissance England* (Baltimore 1937), is a survey of English scientific writing in the sixteenth and early seventeenth centuries which considers the influence of the Copernican and Tychonic theories of the universe on English astronomical ideas.

Walter Pagel has written an important account of *Paracelsus* (Basle and New York 1958), and A. G. Debus is good on *The English Paracelsians* (London 1965). The best biography of Vesalius is by C. D. O'Malley, *Andreas Vesalius of Brussels 1514–1564* (Berkeley 1964). Geoffrey L. Keynes has written *The Life of William Harvey* (Oxford 1966) and Harvey's great achievement is considered by C. Singer in *The Discovery of the Circulation of the Blood* (London 1922). Marie Boas, *Robert Boyle and Seventeenth-Century Chemistry* (Cambridge 1958), discusses the work of the period's most important chemical thinker.

Among the many works which deal with the effects of the scientific revolution on the arts are J. N. Douglas Bush, *Science and English Poetry. A historical sketch, 1590–1950* (New York 1950), *English Literature in the Earlier Seventeenth Century, 1600–1660* (2nd edition, Oxford 1962); Marjorie Nicolson, *Science and Imagination* (New York 1956); H. H. Rhys (ed.), *Seventeenth Century Science and the Arts* (Princeton 1961); and James Sutherland, *English Literature of the Late Seventeenth Century*

(Oxford 1969). Paul Hazard, *The European Mind 1680–1715* (Harmondsworth 1964), discusses the great secularization of thought which took place at the end of the seventeenth and the beginning of the eighteenth centuries, partly as a result of the influence of scientific ideas on the educated classes, while Keith Thomas, *Religion and the Decline of Magic* (London 1971), throws a flood of light on the intellectual attitudes of all ranks of English society in the sixteenth and seventeenth centuries.

The beginnings of the application of statistics to administrative problems are considered in: L. Stone, 'Elizabethan Overseas Trade' in *Economic History Review*, 1949–50; G. N. Clark, *Guide to English Commercial Statistics* (London 1938); B. R. Mitchell, *Abstract of British Historical Statistics* (Cambridge 1962); C. H. Wilson, *England's Apprenticeship 1603–1763* (London 1965); J. H. Plumb, *The Growth of Political Stability in England 1675–1725* (Harmondsworth 1967); and Bertrand Gille, *Les Sources Statistiques de L'Histoire de France des Enquêtes du XVIIe siècle à 1870* (Geneva and Paris 1964). E. Strauss, *Sir William Petty, Portrait of a Genius* (London 1954), discusses the career of the man who did so much to promote the statistical approach in England. C. Wilson, *The Dutch Republic* (London 1968) gives a good account of the stimulus which scientific developments provided for the Dutch economy in the seventeenth century, and E. E. Rich and C. Wilson (eds.), *The Cambridge Economic History of Europe*, volume iv, *The Economy of Expanding Europe in the Sixteenth and Seventeenth Centuries* (Cambridge 1967), contains a valuable discussion by A. R. Hall of the effect of scientific discoveries on the economic life and medical practice of the times.

LIST OF ILLUSTRATIONS

the manuscript of John Banister's *Anatomical Tables* in the Hunterian Museum, Glasgow. By courtesy of the Wellcome Trustees

101 Padua University; engraving, seventeenth century. British Museum, Map Room

2, 103 Dissections of the arteries and the nervous system mounted for use in the Demonstration Theatre, seventeenth century. Royal College of Physicians, London

104 Plan of St Bartholomew's Hospital, London, in 1617; from Sir D'Arcy Power, *A Short History of St Bartholomew's Hospital*, London, 1923, after the *Repertory Book* of the Hospital. Medical College Library, St Bartholomew's Hospital, London

105 William Harvey; painting after an etching attributed to Robert Gaywood, *c.* 1649. National Portrait Gallery, London

106 Experiment to prove the circulation of the blood; engraving from William Harvey's *De Motu Cordis*, 1628

107 Robert Boyle; painting after John Kerseboom, *c.* 1689–90. National Portrait Gallery, London

108 Ambrose Godfrey Hanckwitz's Golden Phoenix Laboratory, Covent Garden, London; engraving from *The Complete Dictionary of Arts and Sciences*, 1764. Photo Ronan Picture Library

109 Astronomy; tapestry, sixteenth century. Sala de los Grecos, Nuevos Museos, El Escorial, Madrid. Photo Mas

110 Bernard le Bovier de Fontenelle; painting by Nicolas de Largillière (1656–1746). Musée des Beaux-Arts, Chartres. Photo Bulloz

111 Descartes's system of vortices; engraving from René Descartes's *Principia Philosophiae*, 1644. British Museum

112 René Descartes; engraved frontispiece of René Descartes's *De Homine*, 1662. Photo John R. Freeman and Co.

113 Engraved frontispiece of Thomas Sprat's *History of the Royal Society*, 1667. British Museum, Department of Printed Books

114 Pages from Joseph Glanvill's *Scepsis Scientifica*, 1665. British Museum, Department of Printed Books

115 Woodcut title-page from Bishop John Wilkins's *The Discovery of a World in the Moone*, 1638. British Museum, Department of Printed Books

116 Interior of an art gallery; painting by an unknown Flemish artist, seventeenth century. National Gallery, London

117 John Milton; engraving by William Faithorne, 1670. National Portrait Gallery, London

118 *Sight*; painting by Jan 'Velvet' Brueghel, *c.* 1617. Museo del Prado, Madrid. Photo Mas

119 Engraved title-page from Robert Burton's *The Anatomy of Melancholy*, 1628. British Museum, Department of Printed Books

120 Jonathan Swift; painting by Charles Jervas, *c.* 1718. National Portrait Gallery, London